Neuroethics, Justice and Autonomy: Public Reason in the Cognitive Enhancement Debate

The International Library of Ethics, Law and Technology

VOLUME 19

More information about this series at http://www.springer.com/series/7761

Veljko Dubljević

Neuroethics, Justice and Autonomy: Public Reason in the Cognitive Enhancement Debate

Springer

Veljko Dubljević
North Carolina State University
Raleigh, NC, USA

ISSN 1875-0044 ISSN 1875-0036 (electronic)
The International Library of Ethics, Law and Technology
ISBN 978-3-030-13642-0 ISBN 978-3-030-13643-7 (eBook)
https://doi.org/10.1007/978-3-030-13643-7

Library of Congress Control Number: 2019931855

This Springer imprint is published by the registered company Springer Nature Switzerland AG
The registered company address is: Gewerbestrasse 11, 6330 Cham, Switzerland

To Aleksandra

Contents

Introduction

This book is a result of work conducted at several places: the International Centre for Ethics in the Sciences and Humanities in Tübingen, Germany, the Neuroethics Research Group in Montréal, Canada and the North Carolina State University in Raleigh, United States. As such, it represents a developing philosophical position that has been motivated by academic debates which are more or less distant. At first, I am providing an argument in an ongoing debate over the appropriate normative lenses and constraints in the enhancement debate in neuroethics. After that I will focus my attention on the specific issues in the non-medical uses of stimulant drugs and non-invasive brain stimulation devices. Finally, I will conclude this book by dealing with the criticism and responses my arguments generated among neuroethicists and other interested scholars.

Thus, Part I, Conceptual Presuppositions, deals with what I think are the most important normative issues. First of all, I argue that the scope of the problem dictates which normative issues should carry the most weight. Since the uses of stimulant drugs and stimulation devices for the express purpose of "cognitive enhancement" can be viewed on at least three different levels of decision making (personal, group or non-public, and society-wide or public), I offer what I think is a plausible argument that the public level of decision making takes precedence over other levels and that it at the same time forces the strictures of Public Reason (see Rawls 2001), and makes the normative principles of justice and autonomy most salient.

In Chap. 1, I propose a model for regulating the use of cognition enhancement drugs for nontherapeutic purposes. Using the Rawlsian method of reflective equilibrium (Rawls 1999), I start from the considered judgment of many citizens that treatments are obligatory and permissible while enhancements are not, and with the application of general principles of justice explain why this might be the case. My analysis focuses on, and refutes, three reasons that some influential authors in the field of neuroethics might have for downplaying the importance of justice: (1) Justice applies only to public funds and state action—not to individual choice or corporate actors; (2) "Performance enhancement" does imply questions of justice, while "performance maintenance" does not; and (3) There is no sufficient difference between cognition-enhancement drugs and other technologies to warrant the importance of justice for the debate. The challenges are refuted by taking into account the difference between consumption and tool use, and the influence of socioeconomic pressure for widespread use that existing drugs could have on the basic structure of society and equal autonomy of citizens. The analysis of requirements of justice points to a conclusion that introducing economic disincentives for the use of cognition-enhancement drugs would be the most legitimate public policy. With the imposition of taxes, fees, and requirements of additional insurance, the use and indirect coercion to use would be less profitable and less widespread, while additional funds thus created could be allocated to meet basic medical needs and/or education.

There are several conceptual presuppositions that my analysis of justice assumes: that pressure to enhance is a real problem and that a proposal solution (i.e., a regulatory model) can, in fact, be effective. On pain of being flat out contradicted on these assumptions, additional analysis must be provided before the normative force of the argument can be judged to be publicly convincing.

Therefore, in Chap. 2, I specifically examine the claims in the debate on cognitive enhancement in neuroethics that society-wide pressure to enhance can be expected in the near future. I use rational choice modeling to test these claims and proceed with the analysis of proposed types of solutions. The discourage use, laissez-faire and prohibition types of policy are scrutinized for effectiveness, legitimacy and associated costs. Special attention is given to the moderately liberal discourage use policy (and the gatekeeper and taxation approaches within this framework), as many authors presuppose that this type of policy would best serve public interest. Different more or less articulated models in the taxation approach (Tobacco regulation analogy, Coffee-shop system, Regulatory Authority for Cognitive Enhancements and Economic Disincentives Model) are analyzed from the point of view of justificatory liberalism. I conclude that blanket prohibition and laissez-faire types of policy would neither be effective nor justified. A moderately liberal public policy shows more promise, but not all approaches within this type of policy would be legitimate and effective. The "gatekeeper" approach and related models could not be justified whereas approaches based on taxation with suitable models might be legitimate and effective.

Again, as is often the case in philosophy, new solutions open up new questions and challenges—or make the old ones more salient. There are 'loose ends' in terms of conceptual presuppositions that need to be analyzed in order to proceed with the empirical findings. The most important 'loose end' is the charge that pressure or no pressure to enhance, we really don't need to care since the principle of autonomy does not have any normative force in neuroethics. This is allegedly due to the fact that neuroethics takes neuroscience seriously, and neuroscience shows that autonomy is based on illusions.

Consequently, in Chap. 3, I examine and refute the claims that neuroscientific evidence renders autonomy "quixotic" and thus supports a shift toward paternalism in medical and political decision-making. My argument is that the notion of autonomy has been mistakenly associated with the metaphysical concept of free will, and I offer a political definition of autonomy to clarify how responsibility is implicitly grounded in the legal and political system: An agent acts autonomously when she/he (a) endorses decisions and acts in accord with internal motivational states, (b) shows commitment to them in the absence of undue coercion and compulsion, and (c) could, as a reasonable and rational person, continue to do so after a period of informed critical reflection. I argue further that neuroscientific findings confirm the assumption that humans are fundamentally fallible social creatures and explain the mechanisms of openness to the social world, which can be and sometimes are abused. Scientific findings do not dispute autonomy or rights, but they do point toward the means of manipulation and toward areas in which further legal protection of rights and autonomous choice is needed. I conclude by clarifying the ideal-typical degrees of coercion (indirect, direct and total) and compulsion (mild, severe and total) that serve the purpose of qualifying reduction of autonomy and responsibility in certain cases, and by elaborating the middle-ground position between the "moral" and "brain disease" model of addiction.

In Part II of the book, I focus on the empirical constraints. No amount of conceptual analysis can solve real-world problems if it is not informed by relevant empirical cases. In that sense, a discussion of the phenomenon of cognitive enhancement has to consider the most often used stimulant drugs and brain stimulation devices which are putatively offering cognitive enhancement effects.

To that effect, in Chap. 4, I analyze appropriate public policies for enhancement use of two most important stimulant drugs: Ritalin (methylphenidate) and Adderall (mixed amphetamine salts). I argue that appropriate regulation of cognition enhancement drugs cannot be a result of a general discussion on cognitive enhancements as such, but rather has to be made on a case-by-case basis. Starting from the taxation approach to cognition enhancement drugs, I take another look at available, moderately permissive models of regulation. After a thorough analysis of relevant characteristics of methylphenidate and amphetamine, I conclude that a moderately liberal permissive regulation of enhancement use by healthy adults might be appropriate for extended release forms of methylphenidate. However, due to their danger profile, amphetamine and instant release forms of methylphenidate should not be made readily available to healthy adults and would need to be prohibited.

Having detailed policy proposals for certain types of pharmacological enhancers is certainly a better starting point for further analyses. Namely, newer stimulants and drugs of other types can be subsumed under existing policies, or policies can be tailored to regulate specific challenges posed by either physical effects or social practices associated with such use. However, a different type of use is possible with electromagnetic enhancements. A pill or a tablet of a drug can only be used once—the supply needs to be renewed for continual use, which makes the analyzed policies enforceable. Devices can be reused indefinitely, can be marketed in more ways than one (product, service), and can potentially be built from scratch at home.

In Chap. 5, I analyze two devices that are at the forefront of the cognitive enhancement debate. There is mounting evidence that non-invasive brain stimulation devices—transcranial direct current stimulation (tDCS) and transcranial magnetic stimulation (TMS)—could be used for cognitive enhancement. However, the regulatory environment surrounding such uses of stimulation devices is less clear than for stimulant drugs: a fact that has already been commercially exploited by several companies. Thus, I review the mechanism of action, uses and adverse effects of non-invasive neurostimulation devices, along with specific social and ethical challenges pertaining to their use as cognitive enhancements. Two regulatory approaches that could be used to facilitate responsible use of these devices as products and services are outlined. Apart from establishing the urgently needed comprehensive regulatory framework, they might provide a starting point for establishing long term physiological and social effects of enhancement uses of tDCS and TMS.

Elaborating normative arguments and policy proposals is just the first step. These have to be able to hold their own ground in academic debates and hopefully in public deliberation. Therefore, an important step is to offer a detailed response to the objections raised by the academic community. In the final, third part of the book, I tackle the constructive criticism levelled at the conceptual work (from Part I) and the policy proposals (from Part II).

Therefore, in Chap. 6, I briefly canvass the numerous objections my conceptualization of the political conception of autonomy has garnered. I look at the merits of the counter-arguments and counter-examples and respond by either rejecting the grounds of the critique, by clarifying the stumbling blocks so that critics and I can reach an agreement, or by incorporating the constructive suggestions into conceptual changes. Even as conceptual issues raise debates, they are not less important than empirical issues at the level of policy.

Thus, in Chap. 7, I analyze the considerable amount of constructive criticism regarding my proposal for regulation of Ritalin and Adderall. Some neuroethicists objected to my favoring prohibitive policies to dangerous CE drugs such as amphetamine and argued for laissez-faire or even mandatory use of enhancements. Others took issue with the conclusion that the economic disincentives model (EDM) could be an option for public policy on extended release forms of methylphenidate. Furthermore, there are those that thought my argument in general and EDM in particular are failing to address the relevant issues in regulation of CE, such as social justice and real autonomy. Finally, there are those who offered suggestions on how the argument and the model of public policy for CE drugs can be improved. Again, based on these comments, which are greatly appreciated, I articulated responses which have hopefully captured the controversial issues, and that I hope will propel the public debate forward.

Now, a word of caution for the reader: even though I am convinced by the arguments presented in this book, they are by no means definitive. I hope that the debate captured within these pages will motivate the reader to think critically about the issues of public reason, justice, and autonomy, and I hope they will become engaged in the public discussion on the public policy pertaining to a range of "neuroethically charged" issues. I also hope that there will be many more books in neuroethics, written by authors from different countries (and continents) that force me to reconsider my views.

References

Rawls, J. (1999). *A Theory of Justice—revised edition.* Cambridge: Harvard University Press.
Rawls, J. (2001). *Justice as Fairness: a restatement.* London: Harvard University Press.

Part I
Conceptual Presuppositions: Justice, Autonomy and Public Policy in the Contexts of Neuroenhancement and Neurobiology of Addiction

Chapter 1
The Application of Rawlsian Principles of Justice on Cognition-Enhancement Drugs: A Policy Proposal

Stimulant medication, which has been dubbed "smart drugs," offers the potential for enhancement of cognition, which by itself can be seen as a "promise" or a "threat" due to the drastic changes of the lives of all citizens in society. The current lack of adequate regulation could potentially lead to the widespread violation of rights and justice, especially since direct and indirect coercion may be brought to bear on many individuals as a result of the utility calculations of employers and other corporate actors. The pressure to enhance is likely to become very high in the military and education contexts, but the most far-reaching influence would come from the sphere of business.

Consider the example of logistics companies in a laissez-faire market economy. Let's say that the most profitable trucking routes are fairly long—not shorter than 1250 km (or 775 miles). This long haul run could be achieved in one day, though this would cause considerable stress and fatigue for the driver. Let's say that as things stand (i.e., without enhancement using bio-medical means), trucking companies offer the service of transportation with a duration of 2 days, and the price includes accommodations for the truck driver as a necessity. Let's say that one company decides to assume an employment policy that is preferable to truck drivers who have no problems using stimulant medication, which is used as treatment for attention deficit hyperactivity disorder (ADHD) and narcolepsy, to stay alert and make the run in just one day. This company offers the service for the same price, thus gaining extra profit, but for half the duration. Let's say that the chief competitors of this company respond by offering the "overnight express" service and accordingly give their current employees the following choice: either they will start using stimulant drugs in order to cope with the requirements of the job, or they will be laid off.

This chapter draws and expands on my previous work which has been published as: Dubljević (2012).

V. Dubljević, *Neuroethics, Justice and Autonomy: Public Reason in the Cognitive Enhancement Debate*, The International Library of Ethics, Law and Technology 19,
https://doi.org/10.1007/978-3-030-13643-7_1

The effects on the market are very clear. All long haul trucking companies would either adopt similar policies or go out of business. Most or all of their employees would either consent to using stimulant drugs or be out of work. The choice of both managers and employees is obviously dictated by market forces completely beyond their control. Thus, enhancement technologies, such as stimulant drugs and stimulation devices, could have a profound influence on the everyday lives of most citizens as the working day and deadline expectations change according to the new social and economic pressures.[1]

Having a brief sketch of plausible real-life problems helps clarify why the non-medical use of cognition enhancement drugs (CED) such as stimulants needs to be regulated urgently and adequately. In this chapter, I start with a brief analysis of CED and consider some general implications of Rawls' principles of justice for the moral-political assessment of CED use and legitimate policy options that could resolve potential social problems. Then, I argue that introducing taxes, fees, and requirements of additional insurance (as economic disincentives) for the use of cognitive enhancements would be the most legitimate and responsible policy. After analyzing and refuting some reasons for downplaying the importance of justice in the ethical analysis of CED, I will offer a model for such policy, which would be in accordance with the requirements of justice and protect the autonomy of all citizens—whether they prefer to use cognitive enhancers or not. Finally, I briefly compare this "economic disincentives" model with the "gatekeeper" model of regulation.[2]

1.1 Cognition Enhancement Drugs

Cognitive neuroenhancements in general and CED in particular pose new challenges. On the one hand, they are mind-based enhancements, as opposed to the earlier enhancements of the body, which had much less effect on the mind. Also, unlike genetic enhancements, which would putatively be applied to unborn children, the use of CED concerns competent adults making individual choices for themselves. Moreover, the phenomenon of CED use promises or threatens to change the lives and work of all citizens, not just members of certain professions (athletes, physicians, etc.).

CEDs such as stimulant drugs are easy to produce, administer, and smuggle. Therefore, the cases of healthy adults using these drugs without strict medical need (Glannon 2008) need to be regulated appropriately. Currently available stimulant drugs, such as Ritalin (methylphenidate), Provigil (modafinil) and the more controversial Adderall (mixed amphetamine salts), can undoubtedly provide "performance

[1]It should be noted that this is more likely to happen in countries with weaker protections of workers (e.g., United States, and much of the developing world). However, even countries with better worker protections have had to deal with the issue of employers coercing their employees into stimulant use. For instance, Australia had to include roadside testing for stimulants in order to reduce amphetamine use in truck drivers (see Dubljević 2016 for a longer argument).

[2]The "gatekeeper" approach will be discussed at greater length in Chap. 2.

maintenance" effects; however, "performance enhancement" properties, along with the safety issues, still remain disputed. Performance enhancement means that healthy adults could use these drugs to achieve significantly better results, while performance maintenance merely implies that normal levels of functioning could be maintained while effects of fatigue and sleep deprivation could be reduced (Lieb 2010).

The results of the poll in the journal *Nature* on the use of cognitive enhancement in academia (Maher 2008) and the call for "responsible use of enhancements by healthy adults" (Greely et al. 2008) seem to indicate that pro-enhancement arguments are increasingly viewed as progressive and liberal. But what would the actual application of liberal principles of justice mean for the debate on CED?

The example of trucking companies serves to show that nonexistent regulation could lead to violations of equal rights for citizens who prefer not to use CED. Admittedly, indirect coercion is more often than not associated with the question of autonomy (Appel 2008), but the threats of society-wide violations of equal rights and discrimination are undoubtedly questions of justice. Widespread unregulated use of CED would be unjust because it would undermine the equality of rights and liberties of citizens: those wishing to enhance would have their right to use recognized, while the rights of those who prefer not to use drugs would effectively be diminished. Furthermore, using CED violates the principle of fair equality of opportunity, as the use of drugs might be justified in instances of poor health and clear medical need but not as a means of seeking positional advantage. Having this in mind, one can only wonder why justice has not had a central place in the debate on CED in the field of neuroethics.[3]

1.2 Justice and Legitimate Policy on CED

The common claim of authors opposing enhancement (e.g., Selgelid 2007) is that treatments are obligatory and permissible while enhancements are morally problematic. The application of the principles of justice can explain why this might be the case in CED, once operative definitions of using drugs due to genuine health needs and for cognitive enhancement have been offered.[4] Preventive, curative, rehabilita-

---

[3]In the following, I try to offer an analysis of requirements of justice (as specified in the very influential theory of justice of John Rawls) in the *general case* of CED. Two remarks are in order here: First, a complete assessment of requirements of justice on CED would have to entail a case-by-case basis analysis, which I can deliver only later (in Part II of this book). Second, although I believe that different approaches to justice when applied to the issue of CED would lead to similar conclusions (compare Daniels 2008, 64–78), I cannot pursue this point in this chapter. I can only say that I hope that application of different reasonable conceptions of justice to cognitive enhancements will have a prominent place in the neuroethics debates in the future.

[4]Rawls' principles of justice (in the final formulation) state: "(1) each person has an the same indefeasible claim to a fully adequate scheme of equal basic rights and liberties, which scheme is compatible with the same scheme of liberties for all; (the equal liberty principle); and (2) social and economic inequalities are to satisfy two conditions: First, they are to be attached to positions and offices open to all under conditions of fair equality of opportunity (the principle of fair equality of

tive, and compensatory use of medical drugs is an important part of meeting health needs (Daniels 2008). On the other hand, cognitive enhancement could be defined as use of medical drug(s) for non-health-related improvement of cognition.

Using CED for performance maintenance or enhancement is not an issue of providing basic necessities for those who are lacking, benefiting the least advantaged, or restoring citizens to a position of equal opportunity and liberty, while this is clearly the case in therapeutic uses, when individuals in question are suffering from ADHD or narcolepsy. Furthermore, allowing unfettered access to CED to gain positional advantage could cause erosion to the fabric of society, as citizens would see that medical resources are used as enhancements, while clear cases of disease and impairment are left untreated (Buchanan et al. 2000). This means that justice could be used to draw the line between cases in which it is permissible and obligatory to provide drugs and those in which it is morally problematic and even impermissible. Moreover, as societies' resources are too limited to meet all needs for treatment, justice requires that we meet the most important health needs first. Only if all health needs are taken care of should public funds (i.e., taxpayers' money) be allowed to fund any type of enhancement.

The application of justice has so far only excluded the realistic possibility of having a legitimate claim on public funding for enhancement purposes. But what if private companies and wealthy individuals interested in enhancement provide funding? Would it not be utterly paternalistic to arbitrarily limit their legitimate interests?

CED are presently used by individuals as a means for obtaining undeserved positional advantage (Maher 2008), and the example of truck drivers forced to use stimulant drugs serves to show that CED could be the preferred means of ensuring positional advantage for corporate actors as well. If students use methylphenidate (Ritalin) during an exam because they are diagnosed with ADHD, they are merely being provided with a fair opportunity to compete with other students on equal footing. However, if they use this drug as an enhancement, they are choosing to risk unknown long-term side effects in order to gain advantage over others. Such practices could lead to a situation in which all students need to use CED in order to be competitive.[5] Similarly, all truck drivers would need to use stimulant drugs in order to be able to work at all. Long haul trucking companies would (indirectly) coerce their employees into taking CED in order to gain more profit, while truck drivers would have to take the risks of long-term effects because they are not in a position to refuse. The employees are at the same time robbed of the ability to decide for themselves whether to use enhancements or not and are forced to be the ones bearing consequences of the use. In other words, with the unknown long-term side effects and through coercive practices, CED could create additional disadvantages to those

---

opportunity); and second, they are to be to the greatest benefit of the least advantaged members of society (the difference principle)" (Rawls 2001, 42–43).

[5]Such arguments have already been incorporated in some highly competitive research Universities in the United States. For instance, the honor code of Duke University explicitly forbids "the unauthorized use of prescription medication to enhance academic performance" as a form of cheating (see https://studentaffairs.duke.edu/conduct/z-policies/academic-dishonesty).

who are less affluent and perhaps even lacking basic necessities. Thus, on this interpretation, economic disincentives that would address both individual and corporate behavior are required as a matter of justice.

The principles of justice require that taxes, fees, and requirements of additional health insurance are imposed as economic disincentives for use of CED. Also, the public funds obtained from those who seek advantage by enhancement need to be allocated to the least advantaged. Otherwise, the use of CED by healthy adults would likely maintain or increase rather than reduce social inequality (Glannon 2008). Furthermore, unregulated use of CED could undermine equality in an additional and very important sense: namely, that although there are factual inequalities in the socio-economic and educational status of citizens, they are equal in their ability to formulate and revise their rational life plans, and to have equal opportunity to do so (Rawls 1999). However, if their choice is dictated by market forces—forces that make it economically rational to pursue only one or a limited range of options (such as to use CEDs)—their status as free and equal citizens is undermined. Also, according to the duty of civility, citizens who do wish to enhance should respect the wishes of their fellow citizens not to use drugs and should strive toward public policy that would protect the rights of all.

Principles of justice require, as well, that any medical necessities stemming from the use of enhancements are not financed from public funds, but from private health insurance. This is due to the fact that they are the result of expensive taste (when voluntary) and so should be given the lowest priority in the order of assigning medical resources. Moreover, the principles of justice require that the social pressure to enhance is dealt with efficiently so that no citizens are suffering from coercive practices that force them to take enhancements in order not to lose their jobs. The principle of fair equality of opportunity requires that enhancements be forbidden as a form of cheating in competitive situations. Technically, those who do not use enhancements could be somehow compensated, but it is unclear if any policy could deal with this issue effectively, so prohibition of cheating is the most plausible move. The protection of equal rights and liberty could be achieved by introducing economic disincentives for the use of CED with the imposition of taxes, fees, and requirements of additional insurance, since this would offset any positional advantage. The additional public funds thus generated could be allocated to meet basic health needs of disadvantaged groups and to finance education for talented youth in educationally underserved populations. The pharmaceutical companies earning profits obtained from enhancement drugs could be further taxed or obliged to invest extensively in "orphan drugs"—medications for diseases that are too rare to make it profitable to complete all necessary clinical trials. Such policy would be legitimate, as it is in accordance with the requirements of justice, and it does not undermine the autonomy of citizens any more than is done by taxes on use of other substances which are potentially harmful, such as alcohol and tobacco.

1.3 Challenges to Justice in the Debate on CED

Although the requirements of justice seem to be clear, in the debates on cognitive neuroenhancement some influential authors in neuroethics have downplayed the importance of justice when considering the ethical case and policy options for CED. There could be several more or less plausible reasons for that:

1. Some (e.g., Gazzaniga 2005) might believe that justice applies only to public funds and state action, but does not concern individual choice or corporate actors.
2. Others (e.g., Lieb 2010) might think that performance enhancement does imply questions of justice, while performance maintenance does not.
3. Others still (e.g., Harris 2011; Levy 2007; Savulescu 2006) might argue that there is no sufficient difference between CED and other technologies that would warrant the importance of justice.

So, if justice only applies to public funds and state action, would imposing economic disincentives for CED be fair? Could not a laissez-faire approach be more appropriate?

It is true that citizens and organizations are free to follow their interests as long as they do not harm or defraud anyone, do not violate anyone's rights, and pay taxes. Nevertheless, if these private or partial interests are to be fulfilled at the expense of public interest and/or drastically affect the very basic structure of society (by achieving monopoly, for example), principles of justice apply with full force. Autonomy of citizens and corporate actors would not be violated, since justice is supposed to protect the autonomy and the equal rights of those likely to be adversely affected. The example of long haul trucking companies coercing their employees serves to show that CED could drastically change the basic structure of society (as monopolies and cartels do) and that the state intervention and regulation are not only justified, but also required.

The second challenge is the claim that performance enhancement perhaps does imply questions of justice, while performance maintenance does not. As caffeine, methylphenidate, and modafinil appear to provide only performance maintenance (see Lieb 2010), should we not treat caffeine and other compounds equally? The example of truckers forced to use stimulants is useful yet again. Although coffee is widespread and not regulated (because the effects are well known and modest) and most truck drivers use it, no one is coerced to drink it directly or covertly. It is possible to be a professional driver without drinking coffee, if someone so chooses. A company employing only coffee drinkers or vegetarians (or CED users), though, would be discriminating among citizens according to arbitrary criteria, unduly limiting the freedom of employees, and therefore doing an injustice.[6] Economic disincentives for the use of CED through taxes, fees, and requirements of additional health insurance

[6]Readers more familiar with debates around Rawls' view should keep in mind that my account assumes that cognitive enhancement influences the range of options in the free pursuit of the doctrines of the good life, and this makes the strictures of Public Reason necessary. Furthermore, I draw on Cohen's view of 'pervasiveness of influence' (see Cohen 1997) as the basis for the claim

do not create any injustice, as citizens would be allowed to use them if they so choose and are protected from indirect coercion if they choose not to use drugs.

The final challenge is the claim that there is no sufficient difference between CEDs and other technologies that would warrant the importance of justice. According to this view, injustices stem from the use of any technology, and some groups (e.g., the Amish) are worse off because they choose not to use certain technologies. Nevertheless, that does not give these groups any reason to invoke the issue of justice because they have freely chosen not to use the technologies in question. As for the truck drivers, have they not already accepted the benefits of technology by driving vehicles? Why should they be allowed to invoke justice to stand in the way of progress?

First of all, this challenge does not make the policy of imposing taxes, fees, and additional health insurance any less legitimate. Consider the example of vehicles: in order to use this readily available type of technology, a person must pay *fees* for a training course and pass an exam as proof of driving competence. Then, when the vehicle is bought, *taxes* should be paid. In order to use the vehicle, an appropriate *insurance* must be taken and both the vehicle and the driver should be *registered* by a government agency (subject to additional fees). Finally, while using the vehicle, taxes on fuel, tolls, and appropriate fees for regular technical *inspections* must be paid.

Second, there is a considerable difference between vehicles and stimulant drugs, and *consumption* makes all the difference. Consuming CEDs alters the brain chemistry (e.g., long term balance of neurotransmitters) in ways that are not yet fully understood. Although being happy (e.g., when one gets a new car) or sad (e.g., if one crashes it) changes the brain chemistry as well, these forms of alterations are in balance with background physiological processes, and as such, they are not problematic. Consuming drugs when neurophysiological processes are off balance makes sense (such as in ADHD cases where the lack of neurotransmitter dopamine is causing attention problems), as there is less to be lost, but when the balance is intact, it is rational to be cautious. Citizens who express concerns about the possibility of jeopardizing the balance of neurotransmitters by consumption of CED are clearly reasonable, and not merely opposed to technology, especially since too much dopamine may cause addiction. Therefore, they have every right to invoke justice if economic forces and corporate interests coerce them to consume CEDs or other potentially addictive substances to be able to work.

---

that cognitive enhancement affects the basic structure of society. I am grateful to Tim Hinton for prompting me to make this clear.

1.4 The "Economic Disincentives Model" and Some Possible Objections

After reviewing the objections to using the principles of justice, the proposed model for legitimate regulation needs to be further elaborated and possible objections reviewed. Under the economic disincentives model, a government agency (e.g., Ministry of Health or Food and Drug Administration [FDA], etc.) could offer a licensing procedure to pharmaceutical companies so that drugs that were initially used to treat ADHD or narcolepsy could be marketed under different brand names (and with different quantities of the active substance) for use by healthy adults. These companies would bear the costs of the licensing procedure to put the drug on the market, and they would bear it gladly, as there is considerable amount of money to be made, given the level of interest (see e.g., Maher 2008).

An additional licensing procedure for users would be self-financed by citizens who prefer to be able to have access to CED. In order to use, say, Ritalin, in order to be able to work longer, a citizen should first pay *fees* for a course about known effects and side effects, and pass an exam at a government accredited agency as proof of knowledge. This procedure should suffice as a means of securing *informed* consent to use Ritalin without any medical need. In order to use CED, an appropriate additional medical *insurance* would have to be taken, and the enhancement using citizen should be *registered* and issued a *license* to use this specific drug.

The health of these citizens would be reasonably protected through self-financed obligatory annual medical tests. In effect, CED would be legal, controlled, and monitored—for the time being. The prices of CED would be regulated—they would contain the standard costs of production and distribution, the profit margin would be limited, and an additional tax would be imposed.

Additionally, acting under the influence of CED (without clear medical need) would be impermissible in certain instances, such as examinations or while driving. Moreover, pharmaceutical companies earning profits obtained from CED could be further taxed or obliged to invest extensively in orphan drugs. The public funds gained by taxation of individual users and corporate actors should be invested in providing medical necessities for the least well off, and the remaining funds should be allocated to finance formal schooling for educationally underserved populations.

It could be objected that this model could have considerable drawbacks: the procedures could turn out be expensive, burdensome and bureaucratic, while involving considerable work and time from those regulated.[7] However, this objection fails to undermine the model in any way. For example, the licensing procedure for driving vehicles is bureaucratic and involves considerable time and work from those regulated, and it is still considered efficient and legitimate. Furthermore, the economic disincentives model for CED would function as an additional source and not a further drain for public funds. This policy would efficiently and legitimately ensure that all

[7]The British Medical Association (2007, 34) discusses this drawback for the policy of permissive regulation that is similar to, yet undeveloped and distinct from, the economic disincentives model of regulation. This model will be discussed at length in the next chapter.

citizens could have legal access to CED, and with the imposition of taxes, fees, and requirements of additional health insurance, it would offset any positional advantage from using CED.

Another possible objection could be that the economic disincentives model gives more power to state agencies, and under the principle of subsidiarity, only problems that can't be resolved at lower levels should be passed on for state regulation. For instance, couldn't medical professionals act as "gatekeepers" for CED? Wouldn't such decentralized individual physician-based decision making be better?

First of all, the "gatekeeper" approach would not be as efficient, as it lacks any economic (dis)incentives for the stakeholders. But efficiency is not the only or even the most important issue here—this approach would also lack transparency and legitimacy if applied to CED. Medical doctors have the expertise to diagnose illnesses and prescribe therapy, whereas every citizen should have the right to decide for him- or herself whether to use enhancements or not. The economic disincentives model is more legitimate because it excludes arbitrary decisions as the conditions under which CED can be used are publicly available and negative decisions could be publicly challenged.

One final objection could be that the economic disincentives model would not be successful, as CED would also be available via prescription, so everyone would try to avoid the licensing process. Again, that does not have to be the case, as physicians would support the legitimate model by doing their job responsibly: If there is a publicly recognized procedure for obtaining these drugs, medical professionals would be less inclined to prescribe drugs if they are unsure that these are really needed. In addition, the criteria for prescription could be made strict under guidance from professional and regulatory bodies, and off-label prescriptions should be discouraged.

1.5 Conclusion

The economic disincentives model could be the basis for a reasonable and legitimate public policy on CED. Although longitudinal studies of specific drugs effects need to be taken into account (and the model might be amended accordingly), this policy, taken as a starting point for regulation, would ensure that all citizens could have fair equality of opportunity, and that rights of all citizens are respected. Furthermore, the imposition of licensing procedures, taxes, fees, and requirements of additional health insurance should offset individual and collective positional advantage from using CED.

The example of truck drivers losing their jobs could be successfully tackled with this policy, especially if the prohibition of driving under use was enforced by means of roadside testing. Economic disincentives could curb indirect coercion, so enhancements, even if allowed, would not be as widespread, or at least would not lead to as much injustice.

References

Appel, J.M. 2008. When the boss turns pusher: A proposal for employee protections in the age of cosmetic neurology. *Journal of Medical Ethics* 34 (8): 616–618.

BMA. 2007. *Boosting your brainpower: Ethical aspects of cognitive enhancements*. A discussion paper from the British Medical Association [BMA]. http://www.bma.org.uk/ap.nsf/AttachmentsByTitle/PDFCognitiveEnhancement2007. (accessed March 23, 2012).

Buchanan, A.E., D.W. Brock, N. Daniels, and D. Wikler. 2000. *From chance to choice: Genetics and justice*. Cambridge: Cambridge University Press.

Cohen, G.A. 1997. Where the action is: On the site of distributive justice. *Philosophy & Public Affairs* 26: 3–30.

Daniels, N. 2008. *Just health: Meeting health needs fairly*. Cambridge: Cambridge University Press.

Dubljević, V. 2012. Toward a legitimate public policy on cognition-enhancement drugs. *American Journal of Bioethics—Neuroscience* 3 (3): 29–33. http://dx.doi.org/10.1080/21507740.2012.700681.

Dubljević, V. 2016. Autonomy is political, pragmatic and post-metaphysical: A reply to open peer commentaries on 'Autonomy in Neuroethics'. *AJOB—Neuroscience* 7 (4): W1–W3.

Gazzaniga, M.S. 2005. *The ethical brain*. New York: Dana Press.

Glannon, W. 2008. Psychopharmacological enhancement. *Neuroethics* 1 (1): 45–54.

Greely, H., B. Sahakian, J. Harris, R.C. Kessler, M. Gazzaniga, P. Campbell, and M.J. Farah. 2008. Towards responsible use of cognitive-enhancing drugs by the healthy. *Nature* 456 (7223): 702–705.

Harris, J. 2011. Chemical cognitive enhancement: Is it unfair, unjust, discriminatory, or cheating for healthy adults to use smart drugs? In *Oxford handbook of neuroethics*, ed. J. Illes, 265–272. Oxford: Oxford University Press.

Levy, N. 2007. *Neuroethics: Challenges for the 21st century*. Cambridge: Cambridge University Press.

Lieb, K. 2010. *Hirndoping: Warum wir nicht alles schlucken sollten*. Mannheim: Artemis & Winkler.

Maher, B. 2008. Poll results: Look who's doping. *Nature* 452 (7188): 674–675.

Rawls, J. 1999. *A theory of justice*, revised ed. Cambridge: Harvard University Press.

Rawls, J. 2001. *Justice as fairness: A restatement*. London: Harvard University Press.

Savulescu, J. 2006. Justice, fairness and enhancement. *Annals New York Academy of Sciences* 1093: 321–338.

Selgelid, M.J. 2007. An argument against arguments for enhancement. *Studies in Ethics, Law, and Technology* 1 (1). https://doi.org/10.2202/1941-6008.1008.

Chapter 2
Cognitive Enhancement and the Problem of the Pressure to Enhance: Rational Choice Modeling and Normative Justification

The problem of regulating cognitive neuroenhancement for healthy adults has generated considerable interest, and there is a growing body of literature in neuroethics dealing with this issue (compare Racine 2010, Chap. 6 for an overview). While many authors (e.g. Gazzaniga 2005; Greely et al. 2008) have issued opinions to the effect that such regulation needs to be liberal and permissive, others (e.g. Appel 2008; Capps 2011) have voiced legitimate concerns about the social pressure to enhance if the moderately liberal public policy that seems to be preferred (see Racine 2010, Chap. 6) is too permissive. Although there is no consensus in the literature on the particular approach or model of regulation that should be used, the 'gatekeeper' approach that has been assumed in some influential accounts (e.g. Glannon 2008, 2011; Lieb 2010. However, this 'gatekeeper' approach can be criticized as paternalistic, illegitimate and un-transparent, and I have proposed an alternative taxation approach along with a model which specifies this liberal middle-ground position (in the previous chapter).

In this chapter, I will first analyze whether the assumption about social pressure or indirect coercion is legitimate or ill-conceived by using rational choice modeling. Then, I will shortly discuss the merits of different policy types (prohibition, laissez-faire and discourage use) and focus on approaches within the liberal 'discouraging use' type of policy. I will start this by scrutinizing the gatekeeper approach on grounds of effectiveness and legitimacy—of special interest is the question if this approach is biased toward one part of society at the detriment of specific disadvantaged groups. Several models within the taxation approach (Tobacco analogy, Coffee-shop, Regulatory Authority for Cognitive Enhancements and Economic Disincentives Model) are considered on grounds of effectiveness and whether such policies can be justified to both groups of citizens: those that wish to enhance and those who chose

---

This chapter draws and expands on my previous work which has been published as: Dubljević (2013).

V. Dubljević, *Neuroethics, Justice and Autonomy: Public Reason in the Cognitive Enhancement Debate*, The International Library of Ethics, Law and Technology 19,
https://doi.org/10.1007/978-3-030-13643-7_2

not to. Finally, I will conclude this chapter with brief remarks on the debate about appropriate policy on cognitive enhancement.

2.1 Cognitive Enhancement and Social Pressure

In the neuroethics literature on cognitive enhancement (CE), there are many authors warning about the problem of indirect coercion to enhance (e.g. Greely et al. 2008; Lieb 2010). There is some evidence that in certain parts of society, most notably in higher education and research settings, this has already been the case (see Maher 2008; DeSantis et al. 2008).[1] However, some authors are concerned that CE might have effects in many diverse or even all parts of society. Khushf (2005, 2008), for example, thinks that what he has dubbed "second stage enhancements" will have profound influence through the pressure to enhance in the educational system as a whole, as well as in the military and the economy. There have been some attempts to give more substance to such claims by offering examples from branches of the economy that are rarely linked with cognitive enhancement in the literature. Namely, Appel (2008) examined the pressure to enhance in complex jobs whereas I offered the "truckers forced to take drugs" example (in Chap. 1, and in published articled—see Dubljević 2012a, b) that is supposed to illustrate the profound dangers of allowing corporate actors to pursue positional advantage without regulation even if CE might provide only "performance maintenance."

Whatever the merits of such claims are, they seem to have managed to attract the attention of at least a subset of relevant policy makers. For example, the Science and Technology Options Assessment study for the European Parliament on human enhancement explicitly warns about "second stage enhancements" and their potential to produce society-wide harms through indirect coercion (STOA 2009). In this decade, the impact of CE technologies for the economy and working conditions in the United Kingdom has been addressed by the joint report of the Academy of Medical Sciences, the British Academy, the Royal Academy of Engineering and the Royal Society (AMS 2012).

However, apart from thought experiments, examples, and very limited empirical evidence (that is almost entirely anecdotal), there is little in the way of a sustained rationally constructed argument to make the case that CE could cause enormous pressure to enhance. Using the tools of rational choice theory, such an argument could be provided. The social pressure to enhance could be modeled as a multi-person Prisoners' Dilemma.[2]

---

[1]According to the study by DeSantis et al. (2008), 34% of student participants admitted that they were using ADHD stimulants illegally. Most illegal users reported using ADHD stimulants primarily in periods of high academic stress and found them to reduce fatigue while increasing reading comprehension, interest, cognition, and memory. Furthermore, most had little information about the drugs they used and found procurement to be both easy and stigma-free.

[2]It should be noted here that it is very hard for the modeling strategy to be quantitative, as opposed to qualitative. Namely, currently available data on prevalence varies greatly among different surveys

Let's assume that Alison is assessing her options before an important cognitively demanding task, which could influence the availability of certain options in the future. (I have deliberately left Alison's situation as vague as possible, since this approach should be able to model choices for a diverse class of rational agents in a competitive setting—from students to employees of a corporation.) Her options could be modeled as a game in which Alison is playing against all other competitors.

As can be seen from Display 2.1, the rational decision is influenced by the choices of others, and uncertainty about the use of others leads to using CE being the dominant (or most rational) choice under the circumstances. Furthermore, regardless of whether Alison wants to maximize her maximum payoffs or just maximize the minimum, she would choose to use CE if she is rational. It should be noted here that this analysis did not start with any substantive normative presuppositions about the merits or demerits of CE. Taking CE has not been labeled as cheating nor preferable—the Prisoners' Dilemma has been neutral toward prior preferences. When prior preferences are introduced this analysis also explains the (controversial) finding (Maher 2008) that many people who would otherwise refuse to give CE to their children, would reluctantly do so if other children in the same school are using CE. The conclusion is that if a sufficient number of people expect others to use CE in a given competitive environment that might be enough to start a chain reaction of use of CE so as not to be left out. So far, the warnings about the plausibility of social pressure to enhance seem to be on the point. But what are the proposed solutions? What kind of policies could somehow change this picture and protect different preferences citizens might have? The British Medical Association (BMA 2007, pp. 33–35) proposed a range of regulatory options that branch out from the general prohibition, permissive regulation or 'laissez-faire' free market types of policy. Let's start with the analysis of these general types of policy.

Government non-involvement, or the 'laissez-faire' type of policy has some merits, but mainly it has drawbacks. Namely, although it is universal in reach (i.e. all citizens would have equal access), it is certainly not neutral—as it is the preferred option of the pro-enhancement group. Furthermore, it obviously could lead to the chain reaction of ever more enhancement (as can be seen in Display 2.1), so if the policy on CE is to be fair and justifiable to both groups of citizens—those that want to enhance and those who would rather not—it would have to be something else.

A prohibition type of policy, where the government is very much involved and enforces non-compliance with sanctions, also has some merits and considerable drawbacks. Again, it is universal in reach (i.e. ideally no one would have access), but again it is not neutral—as it is the preferred option of the anti-enhancement group.

---

and ranges from 5 to 35% (see Ragan et al. 2012). Furthermore, the lack of adequate information on long term effects and even short term benefits (the issue how laboratory findings of improvement in cognition relate to every-day performance is far from clear) further complicates the matter. A purely quantitative rational choice modeling strategy would require reliable data, which is not available. Bearing all this in mind, numerical payoffs in the design of dilemmas could only be assigned tentatively and supplanted with a qualitative analysis. This also points toward the conclusion that regulatory models which could provide the missing information would be more effective, even if their preliminary assumptions turn out to be incorrect in the long run.

Display 2.1 A basic multi-person prisoners' dilemma on CE

	Many others do not use CE	Most others use CE
Alison uses CE	[4,1] Alison gains competitive advantage over others	[3,3] Alison gains no advantage, but at least she is not disadvantaged
Alison refrains from using CE	[2,2] Alison does not gain any advantage. She competes on equal footing with those that do not use CE. However, if resources are very scarce (she is competing for number one position), her chances are reduced if at least one competitor is using CE	[1,4] Alison does not gain any advantage, and in fact she is disadvantaged.

Furthermore, it is not clear that it would a) be effective (i.e. really change the chain reaction picture, due to the very plausible possibility of increasing illicit use) and b) that even if it could in principle be effective, in that the costs of enforcement and associated legal harms rained down upon dissenting citizens would be worth taking. Consider first the current situation in Germany, for historical reasons, a bastion of anti-enhancement sentiments. Nominally, prohibition of any type of enhancement is in place: for instance, the possession and use of stimulants such as Methylphenidate (Ritalin) without a prescription is a criminal offense which could be sanctioned by up to three years in prison (Lieb 2010, p. 100). Prescribing stimulants to healthy adults is also a criminal offense, so not only do prescribing physicians risk losing their medical licenses, they are threatened with incarceration as well. And yet according to available data, 33.4% of Methylphenidate (Ritalin) is used off-label, while 12.6% is used without any diagnosis (DAK 2009). Perhaps this ineffectiveness of prohibition could be offset in the future by stricter enforcement and random testing procedures, which are currently not in place.

Let's try to model this situation again with Alison. Let's assume that most others do not use CE (as it is prohibited) and that there is a procedure of random testing for CE (since doping tests on every exam would be too costly). Alison's options should be modeled now as a game in which she is playing against the system.

As it can be seen from Display 2.2, the rational decision depends on the chance of being caught and the severity of sanctions, which pushes the associated costs to ever higher limits. Namely, if tests could be circumvented in any way (e.g. by bringing someone else's clean urine samples) or if sanctions are not high enough, using CE would again be the dominant (or most rational) choice under the circumstances. Therefore, it would also most likely lead to the chain reaction of social pressure to enhance, even though not at the same level as in the 'laissez-faire' case. Furthermore, the prohibition policy itself (and the rising costs associated with it) would have to be justifiable to both citizens that want to enhance and those who would rather not. Thus, it is unsurprising that many authors (e.g. Glannon 2008; Racine 2010) have

Display 2.2 Prisoners' dilemma on CE in a prohibitive environment

	There is no test	There is an appropriate test
Alison uses CE	[4,1] Alison gains competitive advantage over others	[1,4] Alison gets caught. Depending on the sanctions she is either merely disadvantaged or severely limited in available options
Alison refrains from using CE	[2,?] Alison doesn't gain any advantage. She competes on equal footing with those that do not use CE. However, if resources are very scarce (she is competing for number one position), her chances are reduced if at least one competitor is using CE	[3,3] Alison doesn't gain any advantage. She competes on equal footing with others

concluded that a legitimate policy would have to be something else. But what other options are there?

2.2 Liberal Approaches to Regulation—Gatekeeper

In the permissive regulation category, the so called "gatekeeper" approach (relying on health professionals to act as "gatekeepers" of CE) is the most prominent, since it has been assumed in many influential accounts in neuroethics (e.g. Racine 2010; Glannon 2008, 2011). Furthermore, according to one report (Ragan et al. 2012), it is the official policy on CE in Israel. However, it is unclear whether such an approach would solve the problem of social pressure (or just create others) and whether it could be justified to both sides. Additionally, the downsides of the approach have been pointed out, and the whole assumption that the unofficial status quo should remain has come under attack. Among other calls for a responsible policy (e.g., Greely et al. 2008), I have argued that any model within the gatekeeper approach would be an illegitimate and inefficient policy, since it lacks transparency and economic incentives and disincentives for relevant actors (see the previous chapter). Let's try to test this claim by modeling the options in yet another dilemma. Let's say that the society roughly relies on health professionals to act as "gatekeepers" of CE subject to guidance from professional and regulatory bodies (compare BMA 2007, p. 34). Alison's choice is constrained with the decision of a Medical Doctor (MD) (Display 2.3).

It is important to note here that the decision is made by the health professional, which reduces Alison's agency whatever her preferences may be. On the one hand, if Alison's preferred personal choice is not to use CE, she does not consult her MD and is most likely disadvantaged. If she knows that many others are using CE and

Display 2.3 Prisoners' dilemma on CE in a gate-keeper environment

	MD's of many other competitors do not prescribe CE	MD's of many other competitors prescribe CE
Alison's MD prescribes CE	[4,1] Alison gains competitive advantage over others	[?,2] Alison gains no advantage, but at least she is not disadvantaged
Alison's MD refrains from prescribing CE	[2,?] Alison doesn't gain any advantage. She competes on equal footing with those that do not use CE. However, if resources are very scarce (she is competing for number one position), her chances are reduced if at least one competitor is using CE	[1,?] Alison doesn't gain any advantage, and in fact she is disadvantaged

gaining positional advantage, she might reluctantly ask her physician to provide her access to the drugs she needs to compete. If Alison's preferred choice is to use CE she immediately consults a health professional. In any case, the MD makes the relevant decision: if he or she thinks that Alison's particular case is justified, CE will be prescribed, but if not, Alison has the option to go "doctor shopping," until she finds access to CE. Now, that could be circumvented by introducing a model with sterner regulation by the state or regulatory bodies. Perhaps Alison could be limited to only one second opinion. That might resolve the issue of widespread "doctor shopping", but then society is stuck with the issue of unfair access of already privileged members of society. Namely, under the sterner regulation model, MDs would be very careful not to overprescribe CE, while a certain amount of prescriptions would be expected and approved. But which members of society would have access to CE then? It is safe to assume that class differences might have some impact here, so the claims that CE is more likely to increase or maintain social inequality (e.g. Glannon 2008) seem to be on the point. Furthermore, the issues of paternalism and the accumulation of the power to distribute enhancements in the hands of health professionals make justification of this approach to all citizens very hard.

2.3 Liberal Approaches to Regulation—Taxation

I have proposed a taxation approach in discourage use policy (Dubljević 2012a), and elaborated one model that could make this type of policy work in a liberal democratic society (Dubljević 2012b, see also Chap. 1 in this book). However, it is still unclear whether such an approach would solve the problem of social pressure (or just fill the state's budget) and whether it could pass the test of neutrality and be justified to both sides: those that prefer to enhance and those that are opposed to enhancement of any kind. Let's turn to the analysis of several possible models of taxation based 'discourage use' policy.

Discouraging use with taxation could have different forms. One possible form could be similar to the strict tobacco regulation that is currently in force in Norway. The aim of government policy in Norway was to decrease an unhealthy habit which is, in principle, legal. From 1973, when about half the population was smoking, the percentage of use in 2010 has dropped to 19%, which is reasonably successful as 'discourage use' policies go. This has been achieved with anti-smoking measures, such as heavy taxation (up to 200% of the cost) and a ban on the visible display of tobacco products. This created a negative environment for both users and providers. Not least, when the user finally manages to purchase the discouraged product, the package is adorned with graphic images depicting the potential health hazards associated with use, including death from lung cancer. These measures have been designed to create financial burdens and different types of inconveniences for producers, providers and users, that make the decision to use less likely even though it is not forbidden. The rules and regulations in Norway appear to serve as an effective barrier and a legitimate policy for discouraging use (applied to smoking) while not infringing on the autonomy of citizens.

However, it is unclear whether such a model could be equally well suited to CE. Namely, the structure of the user population is certainly different, so similar measures might provide different responses. Norwegians with higher levels of income and education tend to abstain from smoking, or they smoke less. Hard core smokers tend to start earlier, have a lower level of education, live in poorer regions of Norway and earn low incomes. The rational choice modeling assumed decisions related to competitive advantage, whereas tobacco use offers mostly health disadvantages, though admittedly nicotine acts as a mild cognitive enhancer, comparable in strength to tea. Furthermore, taxation of tobacco as a health promotion measure is mainly aimed at encouraging smoking cessation rather than as a barrier to the uptake of smoking. Moreover, most smokers who don't quit (and many who quit but relapse into smoking) just rebalance their home budget to cope with the higher tax rate. Therefore, a similar policy that is more or less effective in the case of tobacco use could potentially be virtually ineffective in the case of CE.

A second option could be to apply a model similar to regulation of so called "soft drugs" in the Netherlands (the "coffeeshop" model in the rest of the text). Soft drugs such as cannabis and hallucinogenic mushrooms are legal for personal use there. As a result, the use of soft drugs (even in public) is not prosecuted. Sale of these drugs, though technically illegal under the still valid Opium Act, is widely tolerated provided that it happens in a limited, controlled way. The legal control of sale regulates designated places (coffeeshops), products (only soft drugs can be sold, not alcohol), quantity (5g maximum transaction), eligible users (only adults, but not limited to citizens in most municipalities), availability of information (a ban on the visible display of the products is in place, and absolutely no advertisement of drugs is allowed), and the political choice of local residents (the local municipality can give the order to close the coffeeshop or to limit access to citizens only).

Again, the rational choice modeling assumed decisions related to competitive advantage, whereas use of soft drugs offers only recreational benefits. Additionally, the use of medical drugs by healthy adults for enhancement of cognitive function has to be dissociated from both therapeutic and recreational uses.[3] Therefore, a similar policy that might be effective in this case could be partially or completely ineffective in the case of CE. Furthermore, an additional problem of "enhancement tourism" might be created with unknown complications. Moreover, even though this regulatory model might make prevalence numbers more clear, it is questionable if it does anything to bridge the gap in the knowledge on short- and long-term effects on the every-day functioning of individuals in society (i.e. apart from laboratory settings).

A third model has been specifically designed for CE. The British Medical Association (BMA 2007) proposed a permissive system of regulation where techniques are permitted under license from a regulatory body—the Regulatory Authority for Cognitive Enhancements (RACE). This rather sketchy proposal suggests that RACE could approve the use of particular CE techniques and issue professional guidance. From the few remarks there are on the model, it could be assumed that it would create financial burdens and inconveniences for producers, providers, and users. However, even BMA envisions drawbacks of such a model: "the establishment of a statutory regulatory body is expensive, bureaucratic and involves considerable work and time from those regulated" (p. 34).

The fourth model, which I proposed, explicitly tackles with the drawbacks of RACE and seeks to limit the costs for society while optimizing regulatory capacities and demands of justification. The Economic Disincentives Model (EDM) has actually been devised as a middle-ground position that could accommodate the interests of both pro-enhancement and anti-enhancement groups in the case of CE drugs. Under this model, an already existing government agency (e.g. FDA or Ministry of Health) would offer a licensing procedure to pharmaceutical companies to market CE drugs for healthy adults. This way, all citizens could have legal access to CE, but with the imposition of taxes, fees and requirements of additional private health insurance, it creates financial and regulatory burdens for their use.

It should be remembered that EDM envisions an additional licensing procedure for users—in order to use CE, citizens would have to pay fees for a course about known effects and side effects and pass an exam as proof of knowledge. Furthermore, an additional medical insurance and obligatory annual medical tests would need to be taken in order to obtain (and renew) a license to use CE. Also, the model would regulate the prices of CE drugs: apart from containing the standard costs of production and distribution, the profit margin would be limited and an additional tax (comparable to those on tobacco products) would be imposed. The companies earning profits obtained from CE would be further taxed unless they assumed the obligation to invest

[3]Bearing this in mind, a prolonged discussion of positions that advocate legalization of "illicit" drugs for recreational use (e.g. De Greiff 1999; Husak 2005) would be misleading, as associated social problems are very different. Namely, there might be peer pressure to consume say cannabis for recreational purposes or as a life-style choice, but pressure to use CE would be felt in entirely different social milieus—competitive work and educational settings.

extensively in orphan drugs. The funds gained by such policy would be invested in providing medical necessities for the least well-off, and any remaining funds would be allocated to finance education for educationally underserved populations, including first generation university students.

Although this model is designed specifically for CE drugs, and thus it might avoid possible problems of tobacco and coffee-shop models of regulation, still it is unclear whether such an approach would solve the problem of indirect coercion and whether it could be justified to both sides. Let's turn to the preliminary analysis of effectiveness of taxation approaches and the related justificatory problem.

2.4 Taxation of CE and Justification in a Democratic Society

The major objection levelled at 'discourage use' approaches is that they already presuppose a normative standpoint that might be prejudicial toward citizens wishing to enhance. Even if there might be health costs associated with the use of CE, as was the case with tobacco use, it is not self-evident that these would be legitimate policies. This point deserves a bit of elaboration. Modern democratic societies are characterized by a plurality of different worldviews, and important positions in political theory have tried to formulate an impartial standpoint that could adjudicate between conflicting claims of such worldviews. Rawls' "Justice as Fairness" (2001) and "Political Liberalism" (2005) are at the forefront of these attempts.

That is why I have started from the analysis of Rawls' principles of justice in the context of CE in order to justify the general taxation approach and the Economic Disincentives Model (see Dubljević 2012a, b and Chap. 1 in this book). However, Rawls' principles of justice are far from being uncontroversial.[4] The difference principle, for example, has come under heavy attack from almost all major political theorists from the '70s onwards (e.g. Nozick 1974; Miller 1999). If the taxation approach is based on the difference principle (and in the available formulation it explicitly is), then the taxation approach could be suspect from the start. However, a legitimate public policy on CE need not be based on a single disputed theory of justice to be justified to citizens. Gaus (2011) has formulated an influential account of justificatory liberalism that avoids the quandaries of resolving such hard issues such as the exact principles of justice which should guide decision making. Namely, if a rule or policy could be presented to citizens as rational within their own evaluative framework, then it is justified by neutral means and does not need support by additional substantive principles or theories. This should not be understood as denying the relevance of principles of justice, but merely as extending the normative force of the argument to those that are critical of any Rawlsian view.

---

[4]Although these principles have had a great impact in bioethics (see e.g. Daniels 2008), there are dissenting voices on applicability of justice to bio-medical issues (e.g. Gert et al. 2006).

Display 2.4 Prisoners' dilemma on CE in a taxation environment

	Regulation (some sort of taxation approach) is in place so many others do not use CE	There is no regulation and most others use CE
Alison (or Betty) uses CE	[3(4),3] She gains competitive advantage, but pays the costs for it (financial and health)	[2(2),2] She gains no advantage, but pays the costs (financial and health related)
Alison (or Betty) refrains from using CE	[3(2),3] She doesn't gain any advantage but she doesn't pay health costs and advantage of competitors is not significant as they pay financial and health costs	[1(1),4] She doesn't gain any advantage, and in fact she is disadvantaged, but she doesn't pay any health costs

However, some citizens in democratic societies might be skeptical about any kind of *liberalism* while endorsing democracy. Fortunately, since there are similar solutions coming from a "radical-democratic" line of thought, this additional line of justification might be sufficient. Namely, Habermas (2005a, b) offered the principle of universalization that stresses the requirements of a democratic procedure for justification of rules or norms:

> A norm is valid when the foreseeable consequences and side-effects of its general observance for the interests and value-orientations of each individual could be jointly accepted by all concerned without coercion (Habermas 2005a, p. 42).

So, could a taxation approach be equally acceptable to both sides and in the best interest of everyone? Let's assume that Alison would not like to use CE as a personal preference, whereas Betty would. Their choices can again be modeled as a game in which each agent is playing against all others.

Now someone could object that regulation policy would be justified only from Alison's point of view, and Alison prefers not to use CE if she can do that and avoid being disadvantaged. But that is clearly not the case. It has been assumed that Betty actually wants to use CE as a matter of personal preference. She has weighed advantages and health costs and she thinks the health costs are a reasonable trade-off to achieve her goals. But notice (on Display 2.4) that in case there is no regulation, Betty doesn't get the advantage (which is the reason she wants to use CE in the first place) but merely endures the health-related costs. Clearly, regulation is rational, in the best interest of both Alison and Betty, and actually provides a legitimate framework in which both can follow their personal preferences.

Another point should perhaps be made here. Even though all four policies in the taxation approach (Tobacco analogy, Coffee-shop model, RACE and EDM) might be justified from the perspective of justificatory liberalism, the requirement of acceptability of foreseeable consequences and side-effects in the principle of universalization clearly favors EDM. Recall that EDM requires that citizens attend a course about known effects and side effects and pass an exam as proof of knowledge. Since according to available data (e.g. DeSantis et al. 2008), most CE users have little information about the drugs they use, and data on prevalence and long term effects is unreliable, the regulatory models have to take into account these problems in order to be effective. The proposal for a Regulatory Authority for Cognitive Enhancements (RACE), even though it is sketchy, might be construed so as to tackle these issues, and the Economic Disincentives Model already envisions requirements that address all of these problems. Therefore, this model might be the most effective policy because it is acceptable to both pro-enhancement and anti-enhancement groups and legitimate from both liberal and radical-democratic points of view. Furthermore, EDM (and perhaps RACE, if amended accordingly) could provide reliable data on consumption and demand, as well as on long-term health costs to CE users. Thus, the effectiveness of public policy could be quantified and revised if necessary.

2.5 Conclusion

The use of CE by healthy adults needs to be regulated and the discussions on this topic should focus less on abstract positions and more on concrete policy proposals. The moderately liberal public policy that seems to be preferred in the literature has to be specified and policy models need to be operationalized. All policy models have to be carefully weighed against costs and benefits (including intended and unintended consequences) and their ability to provide input for evaluation of effectiveness and legitimacy.

The rational choice analysis has shown that CE could in fact create considerable social pressure, and that prohibition and laissez-faire types of policy would neither be effective nor justified. A moderately liberal public policy shows more promise, but not all approaches within this type of policy would be acceptable from the point of view of modern pluralist democracy. In the absence of additional specific information, it appears that the "gatekeeper" approach and related models could not be justified, whereas an approach based on taxation with suitable models might be legitimate and effective. The Economic Disincentives Model (which would allow legal access to CE with the imposition of taxes, fees and requirements of additional health insurance) is the most promising model proposed so far because it can assure state neutrality on personal preferences, protect the best interest of all citizens, provide reliable data on consumption and demand, and promote effective evaluation of long-term health costs among CE users. However, further discussion is needed in order to generate as many proposals for approaches and specific models as possible. CE can be regulated appropriately only as a result of democratic decision making following a public discussion on a sufficiently large eligible set of policy options.

References

Academy of Medical Sciences [AMA]. 2012. Human enhancement and the future of work. Joint report of the Academy of Medical Sciences, the British Academy, the Royal Academy of Engineering and the Royal Society. http://royalsociety.org/uploadedFiles/Royal_Society_Content/policy/projects/human-enhancement/2012-11-06-Human-enhancement.pdf. (accessed November 10, 2012).

Appel, J.M. 2008. When the boss turns pusher: A proposal for employee protections in the age of cosmetic neurology. *Journal of Medical Ethics* 34 (8): 616–618.

BMA. 2007. *Boosting your brainpower: Ethical aspects of cognitive enhancements.* A discussion paper from the British Medical Association [BMA]. http://www.bma.org.uk/ap.nsf/AttachmentsByTitle/PDFCognitiveEnhancement2007. (accessed March 23, 2012).

Capps, B. 2011. Libertarianism, legitimation, and the problems of regulating cognition-enhancing drugs. *Neuroethics* 4 (2): 119–128.

DAK. 2009. Gesundheitsreport 2009. Analyse der Arbeitsunfähigkeitsdaten. *Schwerpunktthema Doping am Arbeitsplatz*. Berlin: IGES.

Daniels, N. 2008. *Just health: Meeting health needs fairly*. Cambridge: Cambridge University Press.

De Greiff, P. (ed.). 1999. *Drugs and the limits of liberalism*. New York: Cornell University Press.

DeSantis, A.D., E.M. Webb, and S.M. Noar. 2008. Illicit use of prescription ADHD medications on a college campus: A multimethodological approach. *Journal of American College Health* 57 (3): 315–324.

Dubljević, V. 2012a. Principles of justice as the basis for public policy on psycho-pharmacological cognitive enhancement. *Law, Innovation and Technology* 4 (1): 67–83.

Dubljević, V. 2012b. Toward a legitimate public policy on cognition-enhancement drugs. *American Journal of Bioethics—Neuroscience* 3 (3): 29–33.

Dubljević, V. 2013. Cognitive Enhancement, Rational Choice and Justification. *Neuroethics* 6 (1): 179–187. https://doi.org/10.1007/s12152-012-9173-5.

Gaus, G. 2011. *The order of public reason: A theory of freedom and morality in a diverse and bounded world*. New York: Cambridge University Press.

Gazzaniga, M.S. 2005. *The ethical brain*. New York: Dana Press.

Gert, B., C.M. Culver, and K.D. Clouser. 2006. *Bioethics—A systematic approach*. Oxford: Oxford University Press.

Glannon, W. 2008. Psychopharmacological enhancement. *Neuroethics* 1 (1): 45–54.

Glannon, W. 2011. *Brain, body and mind: Neuroethics with a human face*. New York: Oxford University Press.

Greely, H., B. Sahakian, J. Harris, R.C. Kessler, M. Gazzaniga, P. Campbell, and M.J. Farah. 2008. Towards responsible use of cognitive-enhancing drugs by the healthy. *Nature* 456 (7223): 702–705.

Habermas, J. 2005a. *The inclusion of the other: Studies in political theory*. Cambridge: Polity Press.

Habermas, J. 2005b. *Justification and application*. Cambridge: Polity Press.

Husak, D. 2005. For drug legalization. In *The legalization of drugs: for and against*, ed. D. Husak and P. De Marneffe, 3–108. New York: Cambridge University Press.

Khushf, G. 2005. The use of emergent technologies for enhancing human performance: Are we prepared to address the ethical and policy issues? *Public Policy and Practice* 4 (2). http://www.ipspr.sc.edu/ejournal/Archives0805.asp. (accessed November 10, 2012).

Khushf, G. 2008. The second stage enhancements. In *Emerging conceptual, ethical and policy issues in bionanotechnology*, ed. F. Jotterand, 203–218. Dodrecht: Springer.

Lieb, K. 2010. *Hirndoping: Warum wir nicht alles schlucken sollten*. Mannheim: Artemis & Winkler.

Maher, B. 2008. Poll results: Look who's doping. *Nature* 452 (7188): 674–675.

Miller, D. 1999. *Principles of social justice*. Cambridge, MA: Harvard University Press.

Nozick, R. 1974. *Anarchy, state, and Utopia*. New York: Basic Books.

Racine, E. 2010. *Pragmatic neuroethics: Improving treatment and understanding of the mind-brain*. Cambridge, MA: MIT Press.

Ragan, C.I., I. Bard, and I. Singh. 2012. What should we do about student use of cognitive enhancers? An analysis of current evidence. *Neuropharmacology* 64: 588–595.
Rawls, J. 2001. *Justice as fairness: A restatement*. London: Harvard University Press.
Rawls, J. 2005. *Political liberalism*, Expanded ed. New York: Columbia University Press.
Science and Technology Options Assessment [STOA]. 2009. *Human enhancement study*. The Hague: Rathenau Institute.

Chapter 3
A Proposal for a Reconstruction of the Concept of Autonomy

In many spheres of scholarship, including bioethics and neuroethics, there seems to be a significant misunderstanding involving the conflation of the metaphysical concept of free will with the moral–political concept of autonomy. Ever since Benjamin Libet published the results of his experiments measuring the timing of a decision to move by using electro-encephalography (Libet 1985), neuroscientific findings have been given a new impetus for metaphysical debates, which have mistakenly spilled over into practical philosophy.[1] Some authors have reacted to this alleged confrontation between scientific and ethical conceptions with the claim that free will is an important illusion (e.g., Wegner 2002). Others (e.g., Greene and Cohen 2004), mistaking autonomy for free will, claimed that we need to reform the legal system, especially the way responsibility is conceived and punishment is meted out, due to the fact that free will is an empirical nonentity. Such approaches have mistakenly affected moral and political philosophy in general and the neuroethics debates (including the debate on cognitive enhancement) in particular. This is because, I think, following Rawls (1974, 1985) it ought be self-evident that practical philosophy should avoid metaphysical disputes and our laws and policies (like most common sense judgments)

---

[1] Of course, there are many philosophers who willingly enter metaphysical discussions and link autonomy with free will. I am skeptical about the possibility that scientific evidence will resolve any *metaphysical* disputes, and believe that a pragmatic approach is implicit in ascriptions of responsibility within the legal and political system, and that such approach should also be taken in ethics, especially bioethics and neuroethics. My aim here is to make these implicit pragmatic assumptions more explicit. Henrik Walter (Walter 1999; Mueller and Walter 2010) has expressed somewhat similar views on autonomy, but he has not dissociated autonomy from metaphysical issues. Due to reasons of space, I limit my discussion to a reconstruction and extension of a Rawlsian political account of autonomy. Although similar approaches might be found elsewhere (e.g., Beauchamp and Childress 2009), Rawls was explicit in dissociating moral–political concepts from metaphysical quandaries.

This chapter draws and expands on my previous work which has been published as: Dubljević (2013).

V. Dubljević, *Neuroethics, Justice and Autonomy: Public Reason in the Cognitive Enhancement Debate*, The International Library of Ethics, Law and Technology 19,
https://doi.org/10.1007/978-3-030-13643-7_3

largely do that. In short, *the concept of autonomy embedded in the legal and political system of pluralistic democratic societies does not presuppose any religious or metaphysical doctrine, and certainly not libertarianism nor compatibilism and/or determinism*. Autonomy should be properly viewed as a political construct within a political conception that can (but need not) be supported by different comprehensive, metaphysical or religious, doctrines. Previously mentioned claims about free will and autonomy have been largely and rightly ignored in political philosophy, as the *freedom of will* debate was (correctly) seen as metaphysical while *liberty* in the political sense of citizens as being free and equal has been construed as political and not metaphysical (Rawls 1985, 2005).

In this decade, autonomy has been examined without recourse to the sweeping claims of metaphysical determinism. In an influential discussion that is both ethically and empirically informed, Felsen and Reiner (2011) concluded that "we need to reconsider the principle of autonomy in order to align our moral values with neuroscientific findings." With this work, Felsen and Reiner have presented a challenge to any further work in neuroethics that is substantively using the normative principle of autonomy. Although the general approach of Felsen and Reiner is laudable, especially as they try not to link their discussion with metaphysical doctrines, it is nevertheless unclear whether their definition of autonomy presupposes or advances metaphysical commitments. Therefore, I wish to revisit that conversation as to whether the political concept of autonomy that could serve as a basis for ascribing responsibility in practical philosophy should be further defined or defended.

I begin by briefly reviewing Felsen and Reiner's arguments and discuss their claims that neuroscientific evidence renders autonomy "quixotic," thus forcing us to reconsider the use of the principle of autonomy in normative arguments in neuroethics. Then, I elaborate on the notion of autonomy that is clearly and distinctly independent from metaphysical debates on free will. This political notion of autonomy is consistent with empirical findings (e.g., Moller et al. 2006) and it can be used to ascribe moral and legal responsibility to all adult human beings who are not suffering from debilitating pathologies or subject to oppressive and constricting conditions. This approach includes an operational definition of autonomy together with a distinction between different ideal typical levels of mitigating factors for qualifying diminished autonomy. Finally, objections to my positions are examined and the frugality of my notion of autonomy for neuroethics is tested with the example of addiction—a major issue for any discussion about the merits or demerits of use of cognitive enhancements (e.g., stimulant drugs).

3.1 External Influences and Autonomy

In their discussion on neuroscience and autonomy, Felsen and Reiner focused on empirical findings that can be interpreted to suggest that our decisions are not as free from undue external influence as the standard model of autonomy requires.[2] In quoting studies and works on "priming effects" (Tulving and Schacter 1990), "framing effects" (Tversky and Kahneman 1981), and the "anchoring effect" (Ariely 2008), Felsen and Reiner claim that we are more open to covert external influences and that the standard model of autonomy cannot accommodate these findings. Furthermore, because physicians, corporations, and other agents can take advantage of tactics for influencing decisions such as framing, branding etc., the authors conclude that the principle of autonomy needs to be reconsidered in the areas of medical decision making, treatment of addiction, marketing, and the broad field of political philosophy.

Although Felsen and Reiner's conclusions (stemming from the data on addiction) seem compelling at first glance, their conclusions about the improbability of autonomy in medical decision making, marketing, and political philosophy are much bolder than the data itself would allow. Contrary to their assumption that autonomy presupposes constant independence from external factors, basic autonomy presupposes only the right to refuse certain influences as undue and the ability to make decisions alone or as an equal member of a group. Agents privately or, in democratic deliberation proceedings, publicly decide which influences are acceptable and which ones are undue in accordance with their exercise of political rights. While the *ideal* of autonomy, or autonomy in the maximal sense, exalts the image of an agent who makes choices uninfluenced by factors that are not rationally chosen or endorsed (and this is something that is advisable for very important decisions), the minimal or basic sense of autonomy requires only that the agent is capable of making rational decisions in accordance with his or her own long-term interests, not that it must always be the case (Levy 2012), especially in cases of relatively trivial or mundane decisions (e.g., will I drink tea or a latte?). Indeed, the political account of autonomy (see Rawls 2005, 72–81) presupposes that individuals are embedded in and open to the social world and that their decisions are in large part influenced by what others want and need.

How, then, does this square with priming, framing, and anchoring effects? If agents are already decisionally constituted in a cooperative framework through fair social practices and norms (i.e., they are citizens of a society, or at least members of a group), it is entirely consistent with their rational life-plans to participate in group/social endeavors and to trust fellow citizens unless there are good reasons not

[2]Felsen and Reiner start from the 'hierarchical account' of autonomy, and analyze three presuppositions of autonomy: (1) consistency with individual's higher order beliefs; (2) rationality; and (3) relative independence from external biases. They conclude that executive control theory and empirical findings on the role of prefrontal cortex in decision making corroborate the first, and that the somatic marker hypothesis and the notion of ecological rationality offer an important addition to the second. For reasons of space, I do not deal with their arguments concerning the first two prerequisites, and limit my analysis to the alleged evidence from cognitive neuroscience that renders autonomy "quixotic." It should be noted here that the discussion with Felsen and Reiner is not my primary motive, but merely an occasion to clarify the political notion of autonomy for neuroethics. Moffatt (2011), Jecker (2011), Walker (2011) have offered important criticism of Felsen and Reiner, with which I am very much in agreement. However, their discussions still leave the connection between autonomy and metaphysical commitments unclear, and I strive to make this aspect clear.

to. It should be fairly uncontroversial to assert that a commitment to conform to the group norm and the unconscious processing of symbolic cues (e.g., priming, framing, etc.) is entirely consistent with this model of cooperation and participatory social practice. Biases (e.g., framing and anchoring effects) in reasoning have already been adequately incorporated in contemporary political philosophy through the ideas of "reasonable pluralism" and "burdens of judgment" (see Rawls 2005, 54–66) . Reasonable citizens[3] in a democratic society tolerate different comprehensive doctrines (including various religions and flavors of metaphysics) and refrain from coercing others into accepting their favorite or "anchored" views precisely because they recognize fallibility due to "burdens of judgment" or various cognitive biases. Thus, on this view and interpretation of available data, empirical results merely confirm the assumption that humans are fundamentally fallible social creatures and explain the mechanisms of openness to the social world, which can be and sometimes are abused. Does a framework based on scientific evidence necessarily reject or challenge autonomy, then? No, but it might enrich the discussion of autonomy by pointing toward means of manipulation and toward areas in which further legal protection of rights and autonomous choice is needed.

Such protections are sometimes already in place while some might need to be instated or strengthened. For instance, in the economic sphere and apart from familiar rights of consumers to change their minds within a set period of time (e.g., money back guarantee), regulatory bodies might need to be formed and enabled to enforce fair market competition. Additionally, new laws and regulations might need to be put in force in order to prevent new "stealth marketing" practices and protect autonomy.

Although consumer and patient rights are important, there is a further question of compatibility between scientific and moral–political points of view that needs to be answered here (as a result of a common misunderstanding that scientific and ethical spheres are antagonistic). Namely, if free will is an illusion (Wegner 2002), shouldn't we entirely reform the legal, political, and medical system, as it bears on agentic decision making including informed consent, in a more utilitarian fashion (Greene and Cohen 2004) and abandon the language of rights entirely? As a preliminary answer to this challenge it is important to emphasize again that the concept of autonomy is merely human self-determination and does not require full-fledged 'free will' (as in uncaused causation) or any other metaphysical basis. But, since historically autonomy has been connected to 'free will,' most notably in the traditional hierarchical account (Frankfurt 1971), then a second question immediately emerges: namely, how is autonomy different from free will?

[3] I use *reasonable* throughout the text of this chapter as equivalent to appealing to reasons, such as standards, values, and principles that can be accepted as relevant by people who are trying to find ways of co-operating with each other on mutually acceptable terms. Compare Rawls (2005, Lecture IX) and Daniels (2008, 103–142).

3.2 The Political Conception of Autonomy

Autonomy is undoubtedly one of the most valued principles in Western democratic societies and perhaps the most important principle of biomedical ethics (Beauchamp and Childress 2013). Autonomous actions could be analyzed in terms of competent choosers who act (1) voluntarily or intentionally (volitional component), (2) with sufficient information and understanding (cognitive component), and (3) without controlling influences that would override already chosen actions (liberty component). These controlling influences can take the forms of either external coercion (e.g., use of force by others) or internal compulsion (e.g., addiction).[4] Although there is considerable overlap between autonomy and free will in the way the notions of compulsion and coercion are commonly used, the metaphysical discussions on free will make speculative leaps that are unnecessary for a viable moral-political notion of autonomy. Autonomy can and should be construed within the horizon of empirically observable phenomena, and no additional assumptions (e.g., that all action is (pre)determined, that there is mental or even uncaused causation, or that determinism and free will are (in)compatible) need to be made. Furthermore, neuroscience (and indeed any type of science) cannot prove or disprove that the world is deterministic (since this is axiomatic, and therefore beyond the reach of empirical corroboration), and there is important evidence that free will (of any kind) is irrelevant for judgments on moral responsibility (see Roskies 2006). Therefore, it is rational and reasonable to assume an agnostic stance and steer clear of metaphysical disputes when ascribing responsibility, or dealing with any moral-political issues. But what capacities are relevant for responsibility if metaphysical debates are dismissed as less irrelevant in this context? I argue that ascriptions of responsibility depend on the three components (volitional, cognitive and liberty) of the moral–political notion of autonomy, as discussed above.

Apart from capacities to form and assess long-term intentions, develop plans, and exert rationality that Reiner and Felsen discuss at greater length (2011, 4–8), autonomy presupposes having mature powers of volition, that is, the capacity of self-control or self-regulation, but not that these need be exercised all the time and for all decisions. Self-control can be depleted and imperfect, but empirical research suggests that it is undoubtedly present in most adults (Baumeister et al. 1998; Moller et al. 2006; Vohs et al. 2008),[5] and prudent adults conserve it for very important

[4]I use the notion of coercion in its classical "enforcement" form. For a good overview of the literature and counterintuitive results of different definitions of coercion see Anderson (2011). A further point has to be made here: Nordenfelt (2007) also uses the notion of compulsion to describe internal motivations of individuals, but he uses compulsion in a strictly negative or welfare-threatening sense in his analysis of psychopathologies. The concept used here is broader and incorporates both compulsions that are not negative (such as consuming oxygen) and those that disrupt rational life-plans (such as addiction to heroin).

[5]Vohs et al. (2008, 884) equate self-control and self-regulation (which is the literal translation of word autonomy) and define it as "the self-exerting control to override a prepotent response, with the assumption that replacing one response with another is done to attain goals and conform to standards." Moller et al. (2006), in their self-determination theory, make the distinction between self-

decisions. However, bearing in mind that a great deal of human action actually conforms to social, environmental, and other external factors, autonomous action can be characterized in the following way:

> An agent acts autonomously when she/he (a) endorses decisions and acts in accord with internal motivational states, (b) shows commitment to them in the absence of undue coercion and compulsion, and (c) could as a reasonable and rational person continue to do so after a period of informed critical reflection.[6]

In what follows, I clarify this characterization of autonomy by examining at length the notions of coercion and compulsion. It is important to note that I use the notion of coercion in its classical "enforcement" form (see Anderson 2010) before Nozick's influential discussion made the so-called "pressure" form dominant, at least in analytical philosophy. That is, Nozick (1969) defined coercion in terms of a successful threat. A great deal of literature on coercion follows Nozick in this regard, which actually leads to problematic conclusions. For example, if I were threatened by a robber and I did not succumb but rather used force to defend myself (or was physically overpowered and then deprived of my valuables), then I was not coerced at all. On the other hand, it should be noted that I use the notion of compulsion exclusively for internal influences.

3.3 Threats to Liberty: Coercion

Let us briefly examine the roles that coercion and compulsion play in the reduction of autonomy. All adult human beings are assumed to be responsible for states of affairs that their bodies have causally initiated (and for omissions, in specific cases) unless it can be proven that they were coerced by an outside force or compelled by an inside force *they could not endorse and incorporate in their long-term rational life-plan after a period of informed critical reflection.* Since self-control can be depleted and comes in degrees, the question of potentially justifying social interventions bearing on coercion or compulsion needs to be addressed.

If *coercion* is carefully considered, at least three ideal types of external influence could be distinguished:

1. Indirect coercion (e.g., social pressure, taxation)— influences through expected utility.
2. Direct coercion (e.g., robbery, legal ban)—sanctions for noncompliance.

---

control (which is ego-depleting) and self-regulation (which is not). I assume that further empirical research will resolve this debate, but regardless of the merits of their definitions, these studies make it clear that the basic political notion of autonomy has firm support in empirical literature.

[6]Walter Glannon provides important criticism to the so called argument from illusion, and offers a similar yet distinct definition that he (unfortunately) links with the compatibilist argument for free will. See Glannon (2007, 2011), respectively. Unlike Glannon, I claim that we need not take any stance on metaphysical disputes in order to ascribe responsibility.

3. Total coercion (e.g., force feeding, straitjacket)—exertion of physical force.[7]

Consider, for example, the case of the "narco-boss" who wants all his underlings to be addicts in order to exert total control over them. At first, he offers incentives for his underlings to try and continue to use drugs. They are given an equivalent of an "employee discount" for personal use, and the narco-boss clearly shows that he favors those who follow his wishes in this respect. However, let us suppose that some of the underlings refuse to use drugs. Then the narco-boss threatens them with severe beatings if they do not comply. Let us say that most underlings succumb to the threat although one underling tries to trick the narco-boss into believing that he uses drugs while he is in fact "clean." But suppose the narco-boss finds out that he has been tricked and, apart from arranging the beatings, he also orders that the underling be forcefully made to take the drugs for an extended period of time. Now it should be fairly obvious that the narco-boss is in fact forcing his own decision on others with increasing levels of pressure, and that treating such behavior as instances of coercion with varying degrees of intensity has merit.

Contrary to this example of the narco-boss as a paragon of illicitly undue influence, coercive intervention by the state can be justified when it *protects and respects the reasonable life-plans of citizens*. Although total coercion is the hardest to justify, it can be admissible as a legitimate reaction of the state apparatus to intolerable or unreasonable threats to the individual or common good. If an individual is totally coerced, physical force has been exerted to cause some behavior, such as swallowing, or outcome, such as presence of a substance in a body of an individual, or an individual is physically constrained. For example, if a person is shot with a sleeping dart while driving and the resulting lack of control over the vehicle causes a major traffic accident, this would qualify as total coercion. In such instances, autonomy is completely lacking and hence no responsibility can be imputed. However, if the sleeping dart were used by police officers to prevent the same person from, say, a suicide bombing attack or deliberately running down innocent pedestrians, this form of total coercion would be justified.

Autonomous choice is not entirely absent in cases of direct rather than total coercion, however, though responsibility might be severely reduced. Individuals who threaten others with force (e.g., robbers) are coercing directly, and the society is certainly justified in prohibiting such actions and enforcing sanctions against them. Legal sanctions connected with such prohibitions are also a form of direct coercion, which is in accordance with autonomy because it protects reasonable and rational life-plans which are endorsable and acceptable under fair terms of social cooperation, after a period of informed reflection from actions and life-plans which are not. As an illustration of this claim, consider the following example: The state may threaten

[7]Norman Daniels (2008, 191–217) treats indirect coercion as quasi-coercion, while Emanuel et al. (2005) name it inducement and argue against conflating inducement and coercion. Although naming these forms of external influence differently might have merits, an important sense of these being instances of exercising outside control would get lost. Furthermore, abuse of state power in undemocratic societies shows that laws can and should be treated as forms of coercion that need to be justified.

to imprison those who threaten others with physical violence such that they cannot retain their liberty, carry out such actions, and/or engage in a life-plan that depends on habitually threatening others. The possibility of a stable system of social cooperation thus depends on the existence of such laws, which prohibit arbitrary (threats of) violence to others. If such laws are valuable, then the threat to incarcerate violent offenders may reduce their liberty with respect to certain sorts of action they would like to perform while also enhancing the liberty of others.[8]

The idea of public autonomy presupposes that all citizens could, as reasonable and rational persons, endorse democratically enacted laws after a period of informed critical reflection. Laws that are based on whims of rulers or the partial interests of the elite are not legitimate and diminish autonomy. However, legitimate laws in democratic societies do not. Citizens who obey legitimate laws due to a genuine wish to uphold the law as well as those who are merely avoiding sanctions are acting autonomously on this reading of autonomy. It could be argued that citizens who merely avoid sanctions are not autonomous as their decision is not "authentic." However, the basic notion of autonomy does not presuppose authenticity; reasons for action based on any kind of values or principles would suffice.

Indirect coercion (when it is not grossly unfair) does not diminish autonomy, at least not in healthy adults who are supposed to be equipped with intellectual and volitional resources to delay gratification and resist short-term temptations. If societies and collective actors are offering incentives or disincentives for some actions (such as to smoke or not to smoke), the autonomous choice is intact, so responsibility for actions can be imputed.[9] Society can regulate all sorts of actions in this way, providing that there are good reasons for such regulation and that fairness is not undermined. For example, the state can regulate potentially unhealthy commodities that are appealing (by themselves or by clever marketing) by introducing disincentives (e.g., taxes) to offset this form of indirect coercion. The rationale is that responses to undue indirect, direct, and total coercion by individuals and corporate actors can justify commensurate responses by the state.[10]

---

[8]This example is loosely connected and adapted from Anderson's (2011) analysis on the relation between stealing and property rights.

[9]Rawls (2005, 472, n) briefly discussed two types of voluntariness: rational and reasonable voluntariness. On this view, incentives and disincentives are never compromising rational voluntariness, whereas they might compromise reasonable voluntariness if the system makes it economically rational to act unfairly, for instance to seriously undercut a segment of the population based on arbitrary criteria. This could be an important point, but due to reasons of space I do not pursue it here.

[10]This "degrees of coercion" approach better explains our intuitions—at least better than the "lesser evil principle" (see Anderson 2011). One of the tests for diminishing responsibility through coercion that was proposed was limiting the responsibility of someone acting under coercion if she or he acts in a way that minimizes the total amount of harm. So, for instance, we might deny that a person is responsible for choosing to injure another to avoid being killed, but hold the person responsible if she or he chooses to kill someone to avoid being injured her- or himself. Though this reading of the lesser evil principle is intuitively plausible, it is not acceptable in non-coercive contexts. For instance, we do not permit one to seize a spare kidney from one individual to save the life of another. The "degrees of coercion" approach does a better job of explaining our intuitions because it presupposes justification through commensurability of *types of coercion*, not of harms.

3.4 Threats to Liberty: Compulsion

At face value, compulsion seems to be a more complex notion than coercion. How can we know if an individual is truly being compelled? Subjective reports alone are insufficient, as many individuals claim they were compelled in order to avoid responsibility and perhaps punishment, for instance, in medical malingering cases. Furthermore, human beings are compelled to breathe, eat, and reproduce, and this should not reduce responsibility in any meaningful way. So, what exactly does compulsion mean in the context of autonomy?

The notion of compulsion presupposes internal influences that are not and could not be endorsed and incorporated in a long-term rational life-plan after a period of informed critical reflection. Again, at least three types of compulsion should be distinguished:

1. Mild compulsion (psychological dependence, reversible).
2. Severe compulsion (physiological dependence, reversible).
3. Total compulsion (physiological dependence, irreversible).

All human beings experience total compulsion to constantly consume oxygen, which most likely is not lamented by any reasonable and rational person. Indeed, it is rational to infer that anaerobic metabolism, even though it could be possible in theory, would require major trade-offs and limit autonomous choices and rational life-plans that individuals hold dear.[11] Therefore, this particular compulsion should not be taken to diminish autonomy or responsibility in any way. The important point here is that in the everyday functioning of individuals, the possibility of rational and reasonable endorsement makes all the difference. If addiction to a certain substance constitutes total compulsion—physiological dependence that is not reversible without total coercion by an outside force (a point I discuss at length in the following)—and reasonable and rational persons could not incorporate this strategy into a long-term rational life-plan, then autonomy has to be considered to be lacking.

Severe compulsions can diminish autonomy and responsibility, but only if they cannot be endorsed and incorporated into a long term rational life-plan under fair terms of social cooperation. For instance, human beings have a compulsion to consume solid food, but this fact does not diminish autonomous choice in any way. It could be possible to survive and perhaps even thrive on fluids—even intravenously—but most people do not find the idea appealing. If addiction to a certain substance constitutes direct compulsion—that is, physiological dependence which is in some cases reversible without outside interference, and in most cases reversible with certain types of outside help—and reasonable and rational persons could not incorporate such use into a long-term rational life-plan, then autonomy is diminished, and responsibility needs to be substantively qualified.

[11] Oxygen is toxic to organisms that have evolved anaerobic metabolism. Even though theoretically humans could be so altered (say, to bear the rigors of space travel), then they would be unable to appreciate the environment for which humans have evolved or participate in even basic activities that many people find valuable, like having a picnic with the family.

Again, since healthy adults are presumed to have sufficient volitional resources, mild compulsion does not diminish autonomy and responsibility. For example, human beings have a mild compulsion to reproduce, but most can employ self-control in different ways to avoid reproduction if they so choose. Similarly, many substances cause mild compulsion—psychological dependence that is in many cases reversible without outside interference. Just as was the case of coercion, the rationale is that mild, severe, and total compulsion in certain cases can justify indirect, direct, and total coercive responses by the state without being paternalistic, i.e., they are protecting and enhancing rather than diminishing liberty.

This reading of autonomy obviously emphasizes the positive role of capacity for reasoning: formulating, assessing, criticizing, revising, and endorsing rational plans or intentions that extend into the near and distant future.[12] However, this capacity for reasoning should not be conflated with the claim that conscious control is maintained all the time, or that reasoning has been employed before each and every action or decision. Most decisions might well be made intuitively and/or with the aid of heuristics: rules of thumb that work well most of the time and reduce our cognitive load (Gigerenzer and Gaissmaier 2011; Sunstein 2005). The long-term rational life-plans and intentions could actually be *automatically* (i.e., non-consciously) followed, but if they *could* be endorsed after a period of informed critical reflection there is no reason to treat them as irrational or contrary to autonomy. Consider the example of religion. Most people do not question religious doctrines into which they have been indoctrinated. Nevertheless, some of these doctrines could be and are endorsed by rational and reasonable persons even after a period of informed critical reflection, although some people revise their commitments and change or completely abandon them. This view is fairly established in political philosophy. For instance, Rawls (2005) defined voluntary (and thus autonomous) choice in terms of endorsing a religion, which is "enough from the point of view of the political" (p. 472). The upshot is that as long there are socially guaranteed ways to protect decisions including 'changes of heart' after critical reflection, autonomy of these decision is the default, and very strong reasons need to be marshalled to prove otherwise.

3.5 The General Usefulness of a Political Concept of Autonomy: The Test Case of Addiction

Even if the political concept of autonomy might be useful for political philosophy, how useful—especially in its frugality or "economic" conciseness—is it for neuroethics and for the debates within it? It could be objected here that this 'political'

[12]The capacity for reasoning presupposes "mental time travel" or capacity to think about past, present, and future (and this capacity has been empirically corroborated; see, e.g., Nyberg et al. 2010; Tulvin 2002) and not merely understanding of available affordances (Gibson 1986), or stimulus driven reactions, and so distinguishes the type of beings autonomy can be ascribed to. For instance, dogs are not autonomous, whereas most adult human beings are.

position is too vague and too far removed from relevant issues in bioethics, let alone neuroethics. For example, the "moral model of addiction" resting on personal responsibility and "autonomy" has been contested by neuroscientists with what they think is a more progressive "brain disease model" (Leshner 1997; Volkow and Li 2005) gaining considerable influence in neuroethics debates(see Carter and Hall 2012; Carter et al. 2012).[13] If the political view of autonomy as I have been describing it presupposes this "moral model" of addiction in the analysis of compulsion, my characterization of autonomy will not convince even a generally educated reader, let alone neuroscientists and neuroethicists.

But a political notion of autonomy again does not take sides between these two disputed positions and need not be troubled by either the moral model or the brain disease model of addiction. Both these models seem to assume that autonomy is an all-or-nothing capacity, which is patently false. Some neuroethicists (e.g., Carter and Hall 2012; Levy 2012) have recognized this fact and have proposed the middle-ground position in the context of addiction: an addict may be capable of choice and suffers from diminished autonomy at the same time. According to this position, autonomy is gradual, it comes in degrees, and the elaboration of ideal-typical constraints (as it was canvassed above) serves the purpose of qualifying these degrees in the political notion of autonomy.

Consider the example of Tommy McHugh:[14] he was a heroin addict incarcerated for violent offenses, and his drug addiction has persisted until a cerebral hemorrhage severely altered his personality. After suffering damage to frontal and temporal lobes, he was effectively cured of his addiction, but at the same time, he developed a *compulsive* interest in painting, sculpting, and writing. Unlike his previous condition (addiction), he is committed to and wholeheartedly endorses his current compulsions and claims that "life is 100% better." The idea of authenticity would mislead us here if it were used as a criterion of evaluation because there is no objective or intersubjective criterion to determine whether endorsing a compulsion is authentic or not. However, the idea of a rational life-plan once again clarifies the difference in these two compulsions: addiction to heroin cannot be incorporated into a long-term rational life-plan, whereas compulsive artistic interest can. One of the reasons for that is what Levy (2012) calls "oscillation of preference," which is discussed in the following, but an additional important reason is that such an addiction is so destructive and harmful to oneself and others that the life of the individual in question is bound to be horrendous. It is important to add that rational life-plans are not a stand-alone category—they should always be assessed against the backdrop of fair terms of social cooperation. For instance, while compulsively painting, sculpting, and writing, Tommy McHugh is also meaningfully socially connected in various ways and does not unduly hinder rational life-plans of others. Therefore, compulsive artistic interest could be endorsed by a reasonable and rational person, as it is not destructive for the

[13] It has to be noted here that the brain disease model was at the core of Felsen's and Reiner's (2011) claim that autonomy has to be reconsidered in the context of addiction.

[14] Glannon (2007, 40–42) recapitulated the objective details of this case.

person in question or others, and is acceptable under fair terms of social cooperation. Contrary to that, a compulsion to consume heroin could not be likewise endorsed.

Indeed, as a result of seeking access to drugs, heroin addicts often engage in risky, degrading, and illegal activities, which is not acceptable as a rational life-plan under fair terms of social cooperation. Also, they spend time and effort both in pursuit of their drugs and attempting to stop consuming the drugs. Levy (2012) calls such inconsistent behavior "oscillation of preference." Heroin addicts constantly oscillate between the preference to consume the drug and the preference to be free from the compulsion to use it because they are "ego depleted." That is, assessing rational options and inhibiting affordances and reflexive responses uses cognitive resources of self-control that their addiction depletes (Baumeister et al. 1998). Normally, these resources are sufficient so that healthy adult humans can plan their future actions reliably, without constant fear that they will impulsively do the opposite of the planned action.

Heroin addicts, on the other hand, cannot trust themselves sufficiently to realize long term intentions because their dopaminergic system has been "hijacked" (to use an expression from the brain disease model) by the substance of abuse. The effects of heroin on brain functioning are greater than those of common environmental rewards (e.g., food, sociability), and chronic heroin use produces changes in the higher cortical areas of the brain that impair the addicted individuals' self-control—the capacity to inhibit the desire to use heroin (Carter and Hall 2012). However, addicted individuals retain *some* degree of control over their drug use and *some* degree of autonomy (Carter and Hall 2012; Levy 2012). For example, although they are severely compelled to take the drug, they are not compelled to commit a robbery in order to get access to drugs.

The ideal typical scales that I have proposed in the preceding sections on coercion and compulsion could be useful in qualifying the extent of loss of autonomy and the exact action that is committed under diminished autonomy. Consider the example of a bank robbery: If someone is threatening a bank clerk with a weapon and demanding money, the bank clerk is undoubtedly directly coerced. The bank clerk makes the rational decision that life has more worth than money and succumbs to the coercion. Responsibility might be lost, while autonomy is temporarily diminished, especially if the bank clerk informs the authorities immediately when the direct coercion (i.e., threat of violence) is no longer present. This subsequent action clearly proves that the first action was not endorsed. In such a case, the bank clerk is clearly committed to upholding democratically enacted laws and is not endorsing helping the criminals in pursuit of their illicit activities. The only cases in which autonomy is completely gone are total coercion and total compulsion. Therefore, both the moral model of addiction that claims that addicts are completely autonomous and the brain disease model that claims that addicts have no autonomy are patently false.

What does all this mean for the case of Tommy McHugh and autonomy? First of all, Tommy McHugh's artistic compulsion is endorsable and stable. He is not suffering from oscillation of preference—in fact, he can reliably plan for the future according to his long-term intentions, and these plans are endorsable by others as well. The fact that he cannot change some of his intentions does not mean that they are not autonomous as long as he is committed to them. Similarly, to this, consumption of, say, nicotine, could in fact cause severe compulsion, but unlike heroin, this compulsion (although certainly unhealthy) is not as destructive and does not entirely disrupt all other rational life-plans a person might have. For instance, while smoking, a person can also be meaningfully socially connected in various capacities.

Direct coercion as a blanket response to nicotine use would be a failed and illegitimate social policy, and would most likely just cause a black market for such products. In fact, since most people do not object to indirect coercion (i.e., civil regulation of safe manufacture, taxation, and public use) of only potentially dangerous substances (Morse 2012), this type of response in the case of nicotine is legitimate and it does not undermine autonomy in any way. My analysis of addiction should not be taken as exhaustive—it has only skimmed the surface of this complex problem. But the goal was not to resolve this issue, but to offer a clarification of the political notion of autonomy for neuroethics that would be useful for future analysis and for discussions on public policy regarding potentially addictive cognitive enhancers (e.g., stimulant drugs).

3.6 Conclusion

To sum up: Empirical evidence coming from cognitive neuroscience or other empirical disciplines does not prove that autonomy is an "illusion." Priming, framing, and anchoring effects merely offer insight into the social nature of humans (including the need for conserving cognitive resources). Although these effects make the *ideal* of autonomy hardly attainable, basic autonomy is unchallenged. Furthermore, the notion of autonomy has been mistakenly associated with the metaphysical concept of free will. Since scientific evidence is unlikely to resolve any metaphysical disputes, the political approach to autonomy needed to be spelled out. My aim in this chapter was to make the implicit pragmatic assumptions in ascriptions of responsibility more explicit. Therefore, I have offered a political definition of autonomy and identified three components (volitional, cognitive, and liberty) that are necessary for autonomous choice.

Furthermore, I have clarified the ideal-typical degrees of coercion (indirect, direct, and total) and compulsion (mild, severe, and total) that serve the purpose of qualifying reduction of autonomy and responsibility in certain cases, and elaborating the middle-ground position between the "moral" and "brain disease" models of addiction. This analysis could be useful for neuroethics in case assessments and in discussions about legitimate and effective protection of consumer and patients' rights, treatment of addiction. Most notably, it will be useful for the further analysis regarding the regulation of potentially dangerous and/or addictive cognitive enhancers (e.g., stimulant drugs).

References

Anderson, S. 2010. The enforcement approach to coercion. *Journal of Ethics and Social Philosophy* 5: 1–31.

Anderson, S. 2011. Coercion. In *The Stanford Encyclopedia of Philosophy* (Winter 2011 ed), ed. E.N. Zalta. http://plato.stanford.edu/archives/win2011/entries/coercion/. (accessed November 10, 2012).

Ariely, D. 2008. *Predictably irrational: The hidden forces that shape our decisions*. New York: Harper.

Baumeister, R.F., E. Bratislavsky, M. Muraven, and D.M. Tice. 1998. Ego depletion: Is the active self a limited resource? *Journal of Personality and Social Psychology* 74 (5): 1252–1265.

Beauchamp, T.L., and J. F. Childress. 2009. *Principles of biomedical ethics* (6th ed.). New York, NY: Oxford University Press.

Beauchamp, T.L., and J.F. Childress. 2013. *Principles of biomedical ethics*, 7th ed. New York: Oxford University Press.

Carter, A., and W. Hall. 2012. *Addiction neuroethics: The promises and perils of neuroscience research on addiction*. Cambridge: Cambridge University Press.

Carter, A., W. Hall, and J. Illes (eds.). 2012. *Addiction neuroethics: The ethics of addiction neuroscience research and treatment*. London: Academic Press.

Daniels, N. 2008. *Just health: Meeting health needs fairly*. Cambridge: Cambridge University Press.

Dubljević, V. 2013. Autonomy in neuroethics: Political and not metaphysical. *American Journal of Bioethics—Neuroscience* 4 (4): 44–51. https://doi.org/10.1080/21507740.2013.819390.

Emanuel, E.J., X.E. Currie, and A. Herman. 2005. Undue inducement in clinical research in developing countries: Is it a worry? *Lancet* 366: 336–340.

Felsen, G., and P.B. Reiner. 2011. How the neuroscience of decision making informs our conception of autonomy. *American Journal of Bioethics—Neuroscience* 2 (3): 3–14.

Frankfurt, H. 1971. Freedom of the will and the concept of a person. *The Journal of Philosophy* 68 (1): 5–20.

Gibson, J.J. 1986. *The ecological approach to visual perception*. Hillsdale, NJ: Lawrence Erlbaum Associates.

Gigerenzer, G., and W. Geissmaier. 2011. Heuristic decision making. *Annual Review of Psychology* 62 (1): 451–482.

Glannon, W. 2007. *Bioethics and the brain*. New York: Oxford University Press.

Glannon, W. 2011. *Brain, body and mind: Neuroethics with a human face*. New York: Oxford University Press.

Greene, J., and J. Cohen. 2004. For the Law, Neuroscience changes everything and nothing. *Philosophical Transactions of the Royal Society in London* 395: 1775–1785.

Jecker, N.S. 2011. Toward a new model of autonomy: Lessons from neuroscience. *AJOB Neuroscience* 2 (3): 50–51.

Leshner, A.I. 1997. Addiction is a brain disease, and it matters. *Science* 278: 45–47.

Levy, N. 2012. Autonomy, responsibility and the oscillation of preference. In *Addiction neuroethics: The ethics of addiction neuroscience research and treatment*, ed. A. Carter, W. Hall, and J. Illes, 139–151. London: Academic Press.

Libet, B. 1985. Unconscious cerebral initiative and the role of conscious will in voluntary action. *Behavioral and Brain Sciences* 8: 529–566.

Moffatt, B. 2011. Neurobiology is not destiny. *AJOB Neuroscience* 2 (3): 48–49.

Moller, A.C., E.L. Deci, and R.M. Ryan. 2006. Choice and ego depletion: The moderating role of autonomy. *Personality and Social Psychology Bulletin* 32: 1024–1036.

Morse, S.J. 2012. Legal regulation of addictive substances and addiction. In *Addiction neuroethics: The ethics of addiction neuroscience research and treatment*, ed. A. Carter, W. Hall, and J. Illes, 261–276. London: Academic Press.

Müller, S., and H. Walter. 2010. Reviewing autonomy: Implications of the neurosciences and the free will debate for the principle of respect for the patient's autonomy. *Cambridge Quarterly of Healthcare Ethics* 19: 205–217.
Nordenfelt, L. 2007. *Rationality and compulsion: Applying action theory to psychiatry*. Oxford: Oxford University Press.
Nozick, R. 1969. Coercion. In *Philosophy, science, and method: Essays in honor of Ernest Nagel*, ed. S. Morgenbesser, P. Suppes, and M. White, 440–472. New York: St. Martin's Press.
Nyberg, L., A.S.N. Kim, R. Habib, B. Levine, and E. Tulving. 2010. Consciousness of subjective time in the brain. *Proceedings of the National Academy of Sciences* 107 (51): 22356–22359.
Rawls, J. 1974. Independence of moral theory. *Proceedings and Addresses of the American Philosophical Association* 48: 5–22.
Rawls, J. 1985. Justice as fairness: Political not metaphysical. *Philosophy & Public Affairs* 14 (3): 223–251.
Rawls, J. 2005. *Political liberalism*, Expanded ed. New York: Columbia University Press.
Roskies, A. 2006. Neuroscientific challenges to free will and responsibility. *Trends in Cognitive Science*. https://doi.org/10.1016/j.tics.2006.07.01.
Sunstein, C.R. 2005. Moral heuristics. *Behavioural and Brain Sciences* 28 (4): 531–542.
Tulving, E. 2002. Chronestesia: Conscious awareness of subjective time. In *Principles of frontal lobe function*, ed. D.T. Stuss and R.T. Knight, 311–325. Oxford: Oxford University Press.
Tulving, E., and D.L. Schacter. 1990. Priming and human memory systems. *Science* 247 (4940): 301–306.
Tversky, A., and D. Kahneman. 1981. The framing of decisions and the psychology of choice. *Science* 211 (4481): 453–458.
Vohs, K.D., R.F. Baumeister, B.J. Schmeichel, J.M. Twenge, N.M. Nelson, and D.M. Tice. 2008. Making choices impairs subsequent self-control: A limited-resource account of decision making, self-regulation, and active initiative. *Journal of Personality and Social Psychology* 94 (5): 883–898.
Volkow, N.D., and T.K. Li. 2005. Drugs and alcohol: Treating and preventing abuse, addiction and their medical consequences. *Pharmacology & Therapeutics* 108: 3–17.
Walker, T. 2011. Full autonomy, substantial autonomy, and neuro-science. *AJOB Neuroscience* 2 (3): 56–57.
Walter, H. 1999. *Neurophilosophie der Willensfreiheit*. Paderborn: Mentis.
Wegner, D.M. 2002. *The illusion of conscious will*. Cambridge: MIT Press.

Part II
Empirical Constraints: The Cases of Major Psychopharmacological and Electromagnetic Cognitive Enhancers

Chapter 4
Neuropharmacology, Addiction and Autonomy: A Proposal for Public Policy on Adderall and Ritalin as Pharmacological Enhancements

The use of medical drugs such as Adderall (mixed amphetamine salts) and Ritalin (methylphenidate) by healthy adults for enhancement of cognitive function is a social trend that has gained in momentum (see, e.g., DeSantis et al. 2008; Maher 2008; Ragan et al. 2012), and accordingly has generated a lot of attention in academia (for an overview see Racine 2010, Chap. 6). A group of influential neuroscientists and neuroethicists (Greely et al. 2008) has generated controversy by issuing a call for responsible use of cognition enhancement drugs (CED) by the healthy. Although these authors provide very clear arguments about the need to regulate the use of CED, unfortunately they haven't been as clear as to what kind of regulation would be conducive to "responsible use." The discussions on this topic have, up until now, tended to focus on abstract theoretical positions, while concrete policy proposals and detailed models are scarce.[1]

While the prohibitive response of the state in the case of drug regulation appears to be discredited (Greely et al. 2008),[2] and there seem to be many problems with the laissez-faire approach (Capps 2011, see also Chap. 2), as it is not clear what

---

[1] Most authors argue about what criterion or which ethical standpoint should be used while assessing CED (for an overview see Glannon 2008), and while there are some discussions of concrete policy options, which were briefly mentioned in Chap. 2 (e.g., British Medical Association [BMA] 2007; Dubljević 2012b), the debate is still very much abstract. However, there are some important contributions noting the urgency (e.g., United Nations Office on Drugs and Crime 2007) and difficulties of regulating CED (e.g., Coenen et al. 2011; Greely 2011).

[2] The current prohibitive response of the state even on the issue of illicit drugs such as heroin seems to be discredited—see, e.g., Husak (2005, 2007), Duke and Gross (1993), and De Greif (1999). However, for important dissenting opinions see, e.g., Wilson (2007) and De Marneffe (2005). It has to be emphasized that even the dissenters agree that the current prohibition regime is too harsh and costly, especially in cases of relatively harmless drugs (e.g., cannabis), and recently, many jurisdictions have started changing the default approach to drugs.

This chapter draws and expands on my previous work which has been published as: Dubljević (2013).

V. Dubljević, *Neuroethics, Justice and Autonomy: Public Reason in the Cognitive Enhancement Debate*, The International Library of Ethics, Law and Technology 19,
https://doi.org/10.1007/978-3-030-13643-7_4

exactly should be the moderately liberal public policy that is apparently preferred in the neuroethics literature (e.g., Glannon 2008; Racine 2010). I have proposed that a taxation approach (see Dubljević 2012a) could show promising results concerning this issue, but there are considerable differences between the active substances in Adderall and Ritalin that should be taken into account. Although there is no doubt that the decision whether to use at least some CED could and should be left to personal choice of individuals (e.g., caffeine tablets), that does not mean there should be any sort of blanket public policy on CED, without discriminating relevant differences. For example, Ritalin (or other formulas of methylphenidate) might be generally safe and effective, whereas Adderall (and other forms of amphetamine) might be dangerous if regulated loosely. Therefore, any proposal on public policy on CED should be made in the context of a case-by-case analysis. In this chapter, I examine the cases of these two controversial stimulants, in order to offer guidance for other substances with similar danger profiles.

This chapter starts by revisiting available moderately liberal policy options (discussed in Chap. 2), but this time analyzed from the point of view of principles in bioethics (Beauchamp and Childress 2009). After that, I conduct a thorough analysis of relevant facts about methylphenidate and amphetamine in order to identify important differences that should be taken into account in any regulatory proposal. I conclude the chapter with proposing models of adequate public policies for regulation of methylphenidate and amphetamine, which can serve as a starting point for regulation of newer CEDs as they appear on the market.

4.1 Policy Option for Methylphenidate and Amphetamine

The proponents of enhancement insist that relatively new stimulants used for enhancement (methylphenidate and amphetamine) are similar to very old ones (coffee and tea), and base the argument on the appeal to the fairness of treating like cases alike. Nevertheless, policy options in a democratic society are not limited to laissez-faire as argued for by most pro-enhancement authors (e.g., Sandberg et al. 2011), or to the strictest form of prohibition as opponents (e.g., Kass 2003) would like.[3] There are also options of regulation so that the individual use is encouraged (e.g., via government incentives) or discouraged (e.g., via taxation), or even to make the use mandatory (Blank 2010). The point is that fairness of treating like cases alike depends on defining sufficiently like cases, and that can only be done by drawing on empirical findings on known effects.

However, some policy options can be put in question even by using abstract principles such as autonomy, beneficence, non-maleficence, and justice. Take mandatory

[3]Furthermore, different models of general policy types can and should be proposed and analyzed in the context of a specific case, and some might be legitimate while others, such as blanket prohibition with punitive sanctions for production, sale, possession, and use, might not. I return to this point in the last section of this chapter.

use of substances, for example—as much as prohibition needs to be justified, the same is the case here because respect for autonomy dictates that personal preferences of competent adults should not be easily overridden. Only if the very autonomy itself is in jeopardy could some such policy be justified. Although one might enjoy, say, a cup of coffee once in a while, at least some citizens would certainly object if drinking coffee were mandatory. Similarly, if the government introduced incentives for tea drinkers, there would certainly be objections on the grounds of justice. So this off-hand analysis seems to leave only laissez-faire and the policy of "discouraging use." In cases in which the substance is fairly harmless (e.g., caffeine) the society does not intrude and each person can choose freely whether to use it or not. However, if there are potential health risks, the state can discourage use of a substance (e.g., nicotine) by introducing taxes or similar measures that do not restrict personal choice.[4]

Even though the four policy options have already been discussed in Chap. 2, it is useful to address them again in the specific context of regulation for amphetamine and methylphenidate. It will be remembered that there are at least four different possible models of "discouragement" with taxation: the tobacco model,[5] the coffeeshop model, Regulatory Authority for Cognitive Enhancement (RACE) and Economic Disincentives Model (EDM).

So, for instance, the antismoking measures, such as heavy taxation and a ban on the visible display of regulated products could be basically copied and applied in the case of Adderall and Ritalin.[6] These measures have been designed to create financial burdens and inconveniences for producers, providers, and users, and appear to serve as an effective barrier and a legitimate policy of discouraging use. However, it is unclear whether such a model could be equally well suited to Ritalin (methylphenidate) and Adderall (mixed amphetamine salts). After all, these are medical drugs with some serious known side effects (possible intoxication being one of them), and it could be too permissive to sell them "over the counter," even if users are given sufficient warning. Furthermore, "tobacco-like" taxation could become an irreplaceable source of income for the state, which can lead to a reluctance to ban the

---

[4]It should be noted here that the "gatekeeper" approach for enhancement use by healthy adults, i.e., allowing use to some by prescription, does restrict personal choice. Therefore, this chapter will not return to further analyze this paternalistic, illegitimate, and untransparent type of policy (Dubljević 2012b, see also Chap. 1), which is largely ineffective as a form of regulation (Dubljević 2013, see also Chap. 2). This effectively reduces the legitimate models of the "discourage use" policy to some kind of taxation approach. Furthermore, prescription of methylphenidate and amphetamine as therapy should be dissociated from the regulatory framework for use of these substances by the healthy, which is the topic of this chapter.

[5]Apart from Chap. 2, more information on the tobacco model as it is enforced in Norway can be found at Euromonitor (2011).

[6]An example could help put this into perspective: a pack of 20 cigarettes costs 90 NOK (the equivalent of 16 USD) in Norway, and two-thirds of the price (60 NOK) is taxes (see http://www.newsinenglish.no/2012/05/14/duty-free-tobacco-comes-under-fire). When this is compared with the highest taxation rate in the United States, the one in force in New York, the "heaviness" of taxation in Norway becomes clear: In New York City, a pack of cigarettes costs 11.9 USD and the total tax on a pack of cigarettes is 5.85 USD (see http://www.huffingtonpost.com/sheelah-a-feinberg/bloomberg-tobacco_b_1542965.html).

substance even if evidence on long term use points toward serious risk of detrimental health hazards.

A second model discussed in Chap. 2 was the "coffeeshop model": copying and applying a model similar to regulation of so-called "soft drugs" in the Netherlands.[7] However, the problematic issue of "enhancement tourism" has been raised as a genuine concern. Namely, as a result of this particular policy, the Netherlands has become a sort of a tourist attraction, and not all societies might share such a tolerant outlook on all the effects and side effects of non-citizens traveling to seek enhancement. Additionally, as mentioned above "soft drugs" provide only recreational benefits (and perhaps a "creativity boost" at the expense of other cognitive abilities). Ritalin (methylphenidate) and Adderall (mixed amphetamine salts), on the other hand, are sought after as cognitive (and perhaps motivational) enhancements and means of positional advantage, so with the introduction of the coffee shop model an unfair competitive environment might be created, with unknown complications.

The third (RACE) and fourth (EDM) model have been specifically designed for cognitive enhancements (see Chap. 2 in this book for details). Although these models might avoid possible problems of tobacco and coffee-shop models of regulation identified above (and EDM explicitly tries to resolve the problems with RACE), again it is unclear whether such an approach would be appropriate. Certain parts of provisions of these models might run into a lot of resistance by corporate actors. For instance, the provisions of EDM on price regulation and taxation of companies profiting from CED could be severely opposed by lobbyist representing the pharma industry, and so the model could be introduced in a truncated form (e.g., only licensing for producers and users). Given the effects methylphenidate and amphetamine have on the dopaminergic pathways in the human central nervous system (CNS), these substances could be dangerous if introduced as a legally available commodity for the general populace. Although it has been assumed that methylphenidate is safer than amphetamine, even Ritalin has been "accused" of creating all sorts of physiological and social harmful effects, from addiction to maintaining racial inequality by overmedicating and pacifying youth of minorities (see, e.g., Breggin 2001; Fitzgerald 2009). Therefore, known facts about methylphenidate and amphetamine have to be carefully analyzed, and harms and benefits have to be weighed before concrete policy options are recommended for these stimulant drugs.

[7] Apart from Chap. 2 in this book, more information on this type of regulation can be found in Staatsblad (2002).

4.2 The Effects of Methylphenidate and Amphetamine

Simply put, methylphenidate[8] and amphetamine[9] in all their various formulations and brand names (most notable are Ritalin and Adderall) are stimulant drugs that affect the dopamine (DA) and noradrenaline (NA) receptors in the CNS. However, there are important differences between these two stimulant drugs. Methylphenidate is a DA and NA reuptake inhibitor, which basically means that it amplifies spontaneously released DA and NA in synaptic clefts (the space between two synapses), and thus affects neurotransmission in the brain. This has the effect of increasing attention and concentration of individuals, especially those who have problems with learning, such as people suffering from attention deficit hyperactivity disorder (ADHD).

Amphetamine, on the other hand, not only inhibits reuptake, but also inhibits monoamine oxidase (MAO) enzymes, which are vital for inactivation and breakdown of monoaminergic neurotransmitters (such as DA and NA), and also *reverses* the DA transporter (DAT) action. In fact, the mechanism of reuptake inhibition is achieved by blocking DAT from gradually transporting used neurotransmitters back inside the presynaptic neuron for reuse, whereas reversal of DAT action influences a further excretion of DA and NA. This means that amphetamine is much more effective, since apart from prolonged presence of already available DA and NA in the synaptic cleft it causes additional release (in high quantity) of these neurotransmitters.

Since DA and NA are important for arousal, attention, and vigilance, both Ritalin (methylphenidate) and Adderall (mixed amphetamine salts) can produce the effect of higher neural activation and a state of heightened concentration, along with decreasing the effects of fatigue.

[8]Although Ritalin is by far the most well known form of methylphenidate, a variety of formulations and (generic) brand names exist. Among these, instant-release (Ritalina, Rilatine, Attenta, Medikinet, Metadate, Methylin, Penid, Rubifen, and Focalin), and extended-release formulas (Equasym XL, Medikinet XL, Metadate CD, Ritalin LA, Rubifen SR, Ritalin-SR, Methylin ER, Metadate ER, methylphenidate SR, Concerta, Watson methylphenidate ER, and Teva-Methylphenidate ER-C) should be distinguished due to different abuse potential. In what follows I draw extensively on Iversen (2008). Unless otherwise noted, this is the source of data in this section. I try to keep the discussion as understandable as possible for a generally educated, non-expert reader.

[9]Amphetamines are a very diverse class of drugs. On the one hand, some amphetamines are medical drugs with legitimate health benefits and regulated purity (e.g., Adderall, Adderall XR, Dexedrine, DextroStat). On the other hand, some amphetamines are illicit drugs known by their street names (e.g., speed) with shifting amounts of various substances (see EMCDDA 2010). To complicate matters further, some drugs (such as Captagon) are originally medical drugs acting as *precursors* of amphetamine (i.e., the human body metabolizes the initial substance into amphetamine), which have gained popularity in the underground scene and then moved entirely into illicit traffic (see EMCDDA–Europol 2011). Furthermore, many discussions include methamphetamine and other substances in the class of amphetamines (see, e.g., Freye 2009), which decreases clarity. Methamphetamine has effects different from and greater toxicity than amphetamine, and is not used as a CED, but only for recreational purposes. Generally, the discussion will be limited to medical drugs containing amphetamine (e.g., Adderall) in the strict sense with regulated purity and that are used for enhancement purposes by healthy adults.

Just how effective these drugs are can be seen from the data in Table 4.1.[10]

As can be seen from Table 4.1, both methylphenidate and amphetamine apparently have an increasing effect on cognitive capacities such as attention and concentration, and a more or less decreasing effect on reaction times and effects of fatigue. This means that healthy adults could use Ritalin and Adderall to be able to work longer and more quickly. That might have been good news apart from the fact that active substances in both these drugs have considerable side effects.

Both methylphenidate and amphetamine (by virtue of having a similar chemical structure) could cause serious cardiovascular adverse events and addiction, while the less debilitating (for the individual user) side effects include nervousness, drowsiness, insomnia, and possible adverse effects during pregnancy. The most immediate adverse effect is the increase in blood pressure, which could be dangerous to individuals that suffer from high blood pressure, and may even cause sudden death. These substances are especially dangerous if they are used in high quantities, injected directly into the bloodstream, or inhaled (e.g., crushed into powder and snorted). The standard, oral use (in moderate quantities) of both these drugs is more or less safe.[11] The drug enters the body via the intestinal tract and is gradually released into the bloodstream (while a portion of the substance gets inactivated by the liver). The drug again gradually enters the brain from the bloodstream (across the so-called blood–brain barrier), and produces the desired effect of modulating neurotransmission. However, if administered intravenously or inhaled, the stimulant drug is no longer released slowly and it can create rapid effects (the so-called rush), euphoric effects (so-called high), and psychiatric adverse events. Apart from similar general

[10]The data on effects of amphetamine use with no sleep deprivation from the Table 4.1 should be taken with a dose of caution: One relevant study (Ilieva et al. 2013) did not find any reliable effects of mixed amphetamine salts on a range of cognitive tasks in a fully rested state by 46 Caucasian young healthy adults, while they reported reliable effects on *perceived* enhancement effects. However, they noted that participants are not representative of the general population (in addition to the restricted age range, they met a number of health and lifestyle criteria for inclusion, including never having used stimulants and low use of coffee), and that only a single dose (20 mg) of a single form of amphetamine has been used.

There is reason for caution in the case of methylphenidate as well, especially since meta-analyses are not equivocal even for therapeutic uses. For instance, the consensus on methylphenidate (MPH) was that there isn't enough information out there to draw substantial positive conclusions about how MPH impacts ADHD and other disorders. Each of the studies found at best a moderate amount of heterogeneity, preventing them from conducting some measures of meta-analysis, and many of the studies were so dissimilar as to bar the researchers from being able to draw any kind of conclusion as to the impact of MPH. Most of the studies found a slightly positive outlook on MPH, where it seemed to improve baseline symptoms with little to no severe side effects, though others found that more mild negative side effects were common and there were some stand out cases of severe side effects associated with MPH. In general, more high quality research needs to be done with regards to MPH, as the current body of evidence is not strong enough to support any kind of overwhelmingly positive outlook on the treatment. See Yan et al. (2018), Tarrant et al. (2018), Auvin et al. (2018), Lee et al. (2018), Liu et al. (2018), Liang et al. (2018), and especially Cortese (2018).

[11]This should also be taken with a dose of caution. The safety of these drugs has been established for treating defined conditions under supervision by a medical professional. Safety of a drug for over-the-counter use by healthy adults is something that would have to be further tested.

Table 4.1 Effectiveness of Methylphenidate and Amphetamine in available randomized control trials (RCT) on healthy adults

Substance (dosage)	Number of RTCs	Sleep deprivation	Number of participants	Age	Fatigue	Vigilance/Attention	Reaction times	Subjective assessment
Methylph. (5–40 mg)	6	No	205	18–40	N.R.	+	(−)	0/(+)
Methylph. (5–40 mg)	1	Yes	20	20–31	N.R.	N.R.	N.R.	0/(+)
Amphet. (10–20 mg)	6	No	154	18–44	(−)	++	−	+
Amphet. (20 mg)	6	Yes	331	18–36	−	++	− −	0

Legend N.R. = No results available, 0 = no effect, (+) weak increase, (−) weak decrease, + moderate increase; − moderate decrease; ++ strong increase; − − strong decrease. Adapted from Lieb (2010, pp. 69 and 73)

short-term side effects, the danger profiles of methylphenidate and amphetamine differ, as can be seen in Table 4.2.

The data from Table 4.2 need to be clarified, of course. First, to explain the way the data has been generated: experts in psychiatry, pharmacology, and addiction rated numerous drugs on three major dimensions of harm (physical health effects, potential for dependence, and social harms) using a 4-point scale, with 0 being no risk, 1 some, 2 moderate, and 3 extreme risk (Nutt et al. 2007). The numbers in the table represent mean values from multiple assessments.[12] The potential for intravenous use as a part of the physical harm profile is relevant both primarily (for achieving higher effects of acute toxicity) and for secondary harms (e.g., spreading of blood-borne viruses). Of special interest for the discussion are also the categories of physical and psychological dependence: physical dependence involves increasing tolerance (higher dosage is needed to produce the desired effect), intense craving, and withdrawal reactions when the drug use is stopped. Psychological dependence is characterized by repeated use of drug, but without tolerance or physical symptoms. Some illicit drugs along with tobacco are included in this table, because they can provide benchmarks against which the harms of methylphenidate and amphetamine can be assessed. With the knowledge of risk assessment of various other substances and models of regulation, sufficiently like cases could be defined.

4.3 The Case of Methylphenidate

Methylphenidate, which is mostly known under the brand name Ritalin, is currently used around the world as a medical treatment for ADHD.[13] However, the use of this drug has been spilling over to the population of healthy adults (for instance, students suffering from ADHD frequently share or sell it to their peers; see Bigelow 2006, 820)

---

[12]The data on danger profiles from Table 4.2 should be taken with a dose of caution. Even though there is a lot of overlap between qualitative assessments of harms in the relevant literature (e.g., Bigelow 2006; Iversen 2008; Miller 2002) and institutional documents (e.g., EMCDDA 2010; UN 1971) and quantitative assessment in the table, experts can be biased in favor or against certain substances. However since this is the only available source of quantified values of drug harms, there is no choice but to rely on it, and to advise that further studies (i.e., assessments by different stakeholders) will be needed.

[13]There is some controversy over the effectiveness of methylphenidate for treatment of ADHD. For instance, Tarrant et al. (2018) argue that while the effectiveness of MPH in treatment of ADHD is well established, their systematic review designed to review all randomized control trials for people with intellectual disabilities (ID) to assess the effectiveness of MPH on core ADHD symptoms yielded that MPH may be effective in some but certainly not all children and adolescents who have ADHD and ID. On average, the number of adolescents who responded to MPH capped at 80% (70–80% for non-ID ADHD children and 40–50% for ID ADHD children), and there was a substantial rate of negative side effects in both ID and non-ID groups (12–24%). Also, the reviewers concluded that the overall quality of the evidence to support MPH effectiveness was very poor. In general, Tarrant and colleagues concluded that the support for MPH is mixed. Cortese (2018) also raised concerns about the level of evidence supporting MPH as a treatment for ADHD. He found that the account raised a substantial controversy with regards to MPH usage, raising concerns that there needs to be more,

Table 4.2 The harm profile of Methylphenidate and Amphetamines according to the Multi-Criteria Drug Harm Scale

	Physical harm				Dependence				Social harm			
	Mean	Acute	Chronic	Intra-venous	Mean	Pleasure	Psych. Depen.	Phys. Depen.	Mean	Intox.	Social	Health costs
Heroin	2.78	2.8	2.5	3.0	3.0	3.0	3.0	3.0	2.54	1.6	3.0	3.0
Cocaine	2.33	2.0	2.0	3.0	2.39	3.0	2.8	1.3	2.17	1.8	2.5	2.3
Amphetamine	**1.81**	**1.3**	**1.8**	**2.4**	**1.67**	**2.0**	**1.9**	**1.1**	**1.50**	**1.4**	**1.5**	**1.6**
Tobacco	1.24	0.9	2.9	0	2.21	2.3	2.6	1.8	1.42	0.8	1.1	2.4
Cannabis	0.99	0.9	2.1	0	1.51	1.9	1.7	0.8	1.5	1.7	1.3	1.5
Methylphenidate	**1.32**	**1.2**	**1.3**	**1.6**	**1.25**	**1.4**	**1.3**	**1.0**	**0.97**	**1.1**	**0.8**	**1.1**

Adapted from Nutt et al. (2007)

and it has been challenged even in the area of therapeutic use, due to increasing rates of prescription, at least in Europe (Ragan et al. 2012). The controversy surrounding methylphenidate is fueled by the fact that it is (along with amphetamine) currently on the list of controlled substances of law-enforcement agencies all over the world. In fact, the United Nations 1971 Convention on Psychotropic Substances explicitly lists methylphenidate as a Schedule II drug (dangerous substance with known medical uses). However, many experts (e.g., Nutt et al. 2007) argue that different (medical and illicit) drugs have been classified in schedules haphazardly, due to historical contingencies, and that the real danger profile often does not correspond with the classification. Of course, no regulatory policy is unchangeable, and the 1971 UN Convention recognizes several ways for change.[14]

The pharmaceutical corporation Novartis (the producer of Ritalin) has been funding various "neutral" appeals to get methylphenidate off the list, which only increased the controversy. As a result of that, many specific claims made by the "anti-Ritalin" lobby (e.g., Breggin 2001) have been empirically tested. Of course, the big pharma industry has a vested interest in loosening of the regulation, so the danger profile should be carefully analyzed and studies should be confirmed by independent researchers before any change in current prohibitive policy is allowed. However, by most accounts, the short-term benefits and cost-effectiveness of methylphenidate are well established. Unlike amphetamine, methylphenidate poses only modest risks (Kocianćic et al. 2004).[15] In fact, if the danger profile of methylphenidate from Table 4.2 (physical harm mean 1.32, dependence mean 1.25, and social harm mean 0.97) is compared to that of benchmark substances—heroin (2.78, 3.0, and 2.54), cocaine (2.33, 2.19, and 2.37), tobacco (1.24, 2.21, and 1.42), and cannabis (0.99, 1.51 and 1.50)—it seems plausible to argue that this case is more like cases of tobacco and cannabis, and less like cases of cocaine and heroin, and could be regulated accordingly.

However, there are other aspects that might weigh in favor of prohibition. The use of methylphenidate by the healthy could be a "gateway" to use of other illicit drugs, such as cocaine and heroin. The basic idea is that since methylphenidate stimulates the CNS and the affects the dopaminergic pathways, its use can "open the door" to

---

better quality studies to raise the level of evidence surrounding MPH. Overall, Cortese concluded that the body of research surrounding MPH is weak and needs to be strengthened by an increased number and more streamlined kind of study into the efficacy, effectiveness and tolerability of MPH for ADHD.

[14]According to Articles 29 and 30 of the United Nations 1971 Convention on Psychotropic Substances, every country has the right to denounce the convention entirely or to propose amendments. However, this is not the only way to propose regulatory change. Article 3, Paragraph 2, explicitly states: "If a preparation containing a psychotropic substance other than a substance in Schedule I is compounded in such a way that it presents no, or a negligible, risk of abuse and the substance cannot be recovered by readily applicable means in a quantity liable to abuse, so that the preparation does not give rise to a public health and social problem, the preparation may be exempted from certain of the measures of control provided in this Convention."

[15]Indeed, a recent meta-analysis found little to no evidence that using Methylphenidate in ADHD patients correlated with an increased risk for cardiovascular disease, which is generally a major concern for all stimulant medications. See Liu et al. (2018).

the use of "harder" drugs and so makes their use more likely. Such arguments have historically been used to argue against legalization of cannabis, although this drug is less dangerous than tobacco. The statistical correlation between cannabis use and later use of heroin and cocaine was enough to establish this more remote danger for autonomy and public health (Robins 1980 and Goode 1999, both quoted in Husak 2005). Regardless of the merits and demerits of the "gateway" argument, according to available empirical data there is no such correlation between methylphenidate and "hard drugs" (see Barkley et al. 2003), and so it is safe to assume that Ritalin is not a gateway drug.[16]

Also, unlike tobacco, methylphenidate does not increase the risk of developing cancer in humans (see Walitza et al. 2007), so it seems that some sort of regulatory model from the taxation approaches discussed earlier might be appropriate.[17] Nevertheless, there is a difference in standard oral use and abuse of methylphenidate. Although moderate use might enhance cognitive function, chronic abusive use can lead to tolerance and psychological dependence with varying degrees of abnormal behavior. Although extremely unlikely, mania and psychosis can be caused if methylphenidate is used intravenously or inhaled (indeed, the danger of intravenous use—1.6 in Table 4.2—is the reason why the physical harm mean is above 1).

However, there is a difference between various formulations of methylphenidate. Time-release technology can effectively preclude non-oral use and danger of addiction (Lieb 2010, 96), so extended-release formulas might have a different danger profile than instant release formulas. This assertion needs to be explained: In Table 4.2 the physical harm mean was calculated by adding harm of acute use (over- dose), chronic use, and possibility of intravenous use and dividing by 3. Table 4.2 reported values for instant-release methylphenidate (the physical harm mean is 1.32, since harm factors are 1.2, 1.3, and 1.6, respectively), and the number reflects the fact that methylphenidate can be extracted from instant-release medications that contain it (e.g., Ritalin) and injected or inhaled in order to achieve euphoric effects.

The values are considerably lower if only standard use is available as an option. For example, the physical harm mean of tobacco is 1.24, even though the acute and chronic factors are 0.9 and 2.9. The fact that the intravenous use factor is 0 significantly decreases the danger profile. If the same logic were used on methylphenidate

---

[16]Of course, bearing in mind the vested interests of both the big pharma industry and anti-Ritalin lobbies, such conclusions should never be based on a single study. However, Merkel et al. (2007) report that most empirical studies have the finding that methylphenidate treatment actually decreases the risk of developing substance abuse disorders (four of these are quoted), while others have found no correlation whatsoever (again, four studies are quoted, and among them Barkley et al. 2003). According to Merkel et al. (2007), only one study has found an increased risk, but the results of this study have not been replicated, so the claim that there is no correlation that would support a "gateway" drug argument is fairly uncontroversial.

[17]Actually, Miller (2002) reports that methylphenidate is correlated with lower than normal incidence of cancer. Therefore, the claim that methylphenidate does not increase the risk of developing cancer in humans is fairly uncontroversial. However, this might point toward the conclusion that tobacco is inadequately regulated, and that an analogy between tobacco and methylphenidate regulation would not be appropriate. This objection is tackled in the last section addressing self-harm and risks of abuse.

extended-release formulas, the danger profile would be considerably lower. In this case, harm factors are 1.2, 1.3, and 0 respectively, so the physical harm mean of extended release forms of methylphenidate would be 0.83. Compared to tobacco (1.24) and cannabis (0.99), methylphenidate extended-release formulas appear to be very safe. Hence, prohibition of use by healthy adults as a form of regulation perhaps might be justified in the case of instant-release formulas, but not in the case of formulas for which it could be proven that they cannot be abused.

But what kind of policy would be legitimate for these "safer" formulas of methylphenidate? Actually, based on the discussion so far, all four policies (tobacco analogy, coffee-shop model, RACE, and EDM) might be justified, but not all of them would be legal and legitimate, that is, in accordance with all requirements of the UN Convention of 1971.[18] Article 3 does state that a preparation may be exempted from the current prohibitive regulatory regime if it "is compounded in such a way that it presents no, or a negligible, risk of abuse and the substance cannot be recovered by readily applicable means in a quantity liable to abuse, so that the preparation does not give rise to a public health and social problem" (UN 1971, 4). Since extended-release formulas of methylphenidate (e.g., Ritalin-SR) apparently cannot be recovered by readily applicable means in a quantity liable to abuse, and the preparation in fact does not give rise to a public health and social problem, this makes all previously reviewed taxation approaches more or less appropriate. However, the convention requires even if a preparation is exempted that the following measures are in place: (a) licenses for manufacture (Article 8); (b) statistical records of quantity, date, supplier and recipient (Article 11); (c) prohibition of and restrictions of export and import (Article 13); (d) inspection of manufacturers, distributors and users (Article 15); (e) statistical reports of use, abuse, and commerce for the UN (Article 16); and (f) penal provisions for illicit manufacture and trafficking in the regulated substances (Article 22).

Although they might be legitimate as a policy of an individual state, both the tobacco analogy and the coffeeshop model do not conform to the requirements of the convention, and would require the state that chooses such a policy to denounce the convention (see Article 29) or to try to impose amendments, and both options have considerable drawbacks (ignoring the convention could also be an option for

[18]The United Nations 1971 Convention on Psychotropic Substances is extremely important since it shaped the development of the global drug control regime. States that have signed this convention (i.e., the majority of UN member states) have very similar domestic legal frameworks for regulation of psychotropic substances, due to compliance with the convention. For example, in the United States, the Psychotropic Substances Act of 1978 explicitly ensures compliance with the convention: "It is the intent of the Congress that the amendments made by this Act, together with existing law, will enable the United States to meet all of its obligations under the Convention and that no further legislation will be necessary for that purpose" (21U.S.C.§801a). Furthermore, the UN Convention on Psychotropic Substances of 1971 should not be confused with the UN Single Convention on Narcotic Drugs of 1961, which shapes regulatory frameworks for opiates (including cannabis). The ambiguous language in that particular treaty makes it unclear whether or not it requires criminalization of drug possession for personal use, which is a fact exploited by the Netherlands, one of the signatory states (it will be remembered that the Netherlands uses the coffee-shop model for cannabis). However, the language in the 1971 convention is unambiguous: Individuals can only use Schedule II substances (including methylphenidate and amphetamine) with a special permission.

rare states that have never signed it). The proposal for a Regulatory Authority for Cognitive Enhancements (RACE), even though it is sketchy, might be construed in accordance with the convention, and the Economic Disincentives Model (EDM) envisions all the requirements from the convention, and thus is the most legitimate.

4.4 The Case of Amphetamine

Could the same logic be applied to the regulation of use of extended-release formulas of amphetamine (e.g., Adderall XR) as well? Apparently not, because the 1971 UN Convention in Resolution II warns that amphetamines in all forms are particularly liable to abuse. This is a question not only of historical contingency, but of empirical fact. Recall that amphetamines differ in the effect on the CNS from methylphenidate because they not only inhibit reuptake of DA and NA, but also inhibit monoamine oxidase (MAO) enzymes, which are vital to inactivation and breakdown of monoaminergic neurotransmitters (such as DA and NA), and also reverse the DAT action (Iversen 2008). The influence on MAO alone increases the danger profile of amphetamines: MAO dysfunction is correlated with a number of psychiatric and neurological disorders, such as depression, schizophrenia, substance abuse, and ADHD. Hence, amphetamines can be very effective in helping individuals with too much MAO (as in ADHD) but cause severe psychotic episodes in people with too little MAO (as in schizophrenia). Indeed, even with oral use of larger quantities, amphetamine can cause aggression, impulsivity, manic behavior, and psychotic episodes (Miller 2002).

The fact that amphetamine reverses DAT increases both the therapeutic effects and the danger of addiction. Methylphenidate is only able to extend the time naturally occurring DA and NA remain in the synaptic cleft, whereas amphetamine causes additional excretion of DA and NA. NA increases arousal, but also increases blood pressure, so additional quantities might cause adverse cardiovascular events in people with high blood pressure. But these are just bodily harms—too much DA can literally "hijack" volitional capacities and impair cognitive capacities of an individual (Hyman 2011). If the amount of DA increases rapidly, an intoxicating effect (rush) is achieved, which impairs volitional capacities and might cause aggression. If the amount of DA is steadily high, it produces pleasurable euphoric effects, which can impair cognitive capacities in the short term (by intoxication) and in the long run (by causing chronic conditions of alternating capacity and incapacity). If this effect is sustained for prolonged periods of time (a week or more), it might even produce psychiatric adverse events that are comparable to positive symptoms of schizophrenia. The so-called amphetamine psychosis is a state of heightened emotional arousal, with frightening visual, auditory, and tactile hallucinations and paranoid delusions. Persons affected can be violent and dangerous to self and others.

Amphetamines are often described as having a high abuse potential, which has the danger of causing "extreme psychological dependence" and "severe social disability" (Bigelow 2006, 234). Quantitatively, if the danger profile of amphetamine from

Table 4.2 (physical harm mean 1.81, dependence mean 1.67, and social harm mean 1.50)[19] is compared to that of heroin (2.78, 3.0, and 2.54), cocaine (2.33, 2.19, and 2.37), tobacco (1.24, 2.21, and 1.42), and cannabis (0.99, 1.51, and 1.50), this case is somewhere between the case of tobacco, which is regulated with taxation, and the cases of cocaine and heroin, which are legitimately prohibited. Although, unlike heroin, amphetamine is not likely to cause death even if abused (Singleton et al. 2009), it certainly is a "gateway" drug for harder substances. It is sometimes described as "poor man's cocaine" and poses a significant social problem as the most abused drug in Europe (EMCDDA–Europol 2011).

Admittedly, amphetamines could provide great benefits if used responsibly. However, the threat of irresponsible use and the fact that amphetamines are frequently abused make a prohibitive response more appropriate. When the principle of beneficence is weighed along with the principle of non-maleficence, it is clear that the dangers of amphetamine use clearly outweigh the benefits. However, it could be objected that the principle of autonomy weighs in favor of a permissive approach, even with amphetamine and instant-release forms of methylphenidate. Isn't prohibition of a substance based on self-harm extremely paternalistic? In order to answer this challenge, the notion of autonomy has to be briefly revisited.

4.5 Abuse, Addiction, and Intoxication as Threats to Autonomy

Since we have clarified the acceptability of the criterion of autonomy for the further normative discussion (in Chap. 3), now we can apply it directly in the context of regulation of cognitive enhancement drugs. It will be remembered that autonomy is one of the most valued principles in Western democratic societies and perhaps the most important principle of biomedical ethics (Beauchamp and Childress 2009). Our operative definition of autonomy framed autonomous actions in terms of competent choosers who act (1) voluntarily or intentionally (volitional component), (2) with sufficient information and understanding (cognitive component), and (3) without controlling influences that would determine actions (liberty component). It should be remembered here that these controlling influences can be external (coercion) or internal (compulsion). Hence, all adult human beings are assumed to be responsible for states of affairs their bodies have causally initiated—and those that they did not but could have in cases of negligence—unless it can be proven that they were coerced by an outside force or compelled by an inside force *they could not endorse*

[19]A further point needs to be explained here. Based on data from Table 4.2 it could be objected that amphetamines are not really addictive. However, recall that the physical dependence rating reflects the increasing tolerance (higher dosage is needed to produce the desired effect), intense craving, and withdrawal reactions when the drug use is stopped. Amphetamines do not cause withdrawal reactions, but do cause intense craving and tolerance, so the rating is 1.1. However, the facts that use can be highly pleasurable (2.0) and that use can cause psychological dependence (1.9) make the threat of addiction very real, especially if amphetamines are abused.

and incorporate in their long-term rational life-plan after a period of informed critical reflection. Drug abuse and addiction have very important consequences for cognitive, volitional, and at least one aspect (compulsion) of the liberty component of autonomy, and might diminish responsibility that accompanies legitimate choices by individuals.

For instance, being addicted to drugs effectively precludes individuals from following and realizing long-term rational life plans, so at least some drugs might be in principle legitimately prohibited. In fact, as noted in Chap. 3, addicts, as a result of seeking access to drugs, often engage in risky, degrading, and illegal activities. Furthermore, as a result of drug effects, regular users are often unable to work for a living. It should be noted again that the effects of most "hard drugs" on brain functioning are greater than those of common environmental rewards (e.g., food, sociability), and chronic drug use produces changes in the higher cortical areas of the brain that impair the addicted individuals' self-control (or volitional component of autonomy)—the capacity to inhibit the desire to use drugs (Carter and Hall 2012; Carter et al. 2012). It might be helpful to further unpack the argument about the threat to autonomy here. After all, are not all dangers of drug use cases of harm to self—which should be perfectly acceptable as long as others are not harmed in any way? And for the argument about social acceptability of drug users, are not the stigma and coercion by society the causes of most of their suffering? It would be hypocritical to argue that drug addicts cannot hold a job and be meaningfully connected with other people as a result of their addiction, when the state scorns and marginalizes them and actively coerces them to the fringes of society and criminal activities.

Let's try to tackle this objection by emphasizing the idea of autonomy as an inalienable right. Although a certain amount of risky activities would certainly not be counter to autonomy (whereas it would definitely be counter to autonomy to prohibit all risky activities), and, say, tobacco use might be one of these, they can be dissociated from a class of "intrinsically debilitating activities" (Freeman 1999, 125). Selling oneself to slavery or permanently mutilating one's cognitive and volitional capacities can be legitimately prohibited, as it can be reasonably assumed that such an option is un-endorsable after a period of critical reflection. Now, a word of caution: in order to fully appreciate even irrational voluntary choices of individuals, the prohibited act should not be the voluntary activity that is "intrinsically debilitating"—it should be forcing, participating, or providing means for another to commit an "intrinsically debilitating activity" (and sanctions should be commensurate to the offense in question). A few examples might be helpful here. Let's say that reasonable and rational people would not endorse a system in which it was possible to sign a contract according to which debtors could be sold to slavery. However, let's say that an individual in really desperate financial circumstances does precisely that—in that they approach an individual or institution and offer to sign such a contract for a sum of money. The society would only be justified in prohibiting the enforcement of such a contract and punishing (with varying degrees of severity) the individual or institution that wanted to benefit from or force others into such a scheme—not the individual in desperate circumstances. Similarly, if an individual for whatever reason does voluntarily and autonomously choose to consume illicit drugs with full knowledge of their addictive properties and harmful physiological and social consequences, the society would be legitimate in punishing the producers and distributors of illicit drugs, while drug addicts might need to be treated and not punished.

The danger profiles have shown that abuse of both methylphenidate and amphetamine, and even oral use of the latter, can lead to a disturbance of a whole range of cognitive, affective, sensory, and volitional capacities. Furthermore, apart from permanent impairment, chronic conditions of alternating capacity and incapacity, if they are likely to produce harms to others (in this case psychosis and mania), could be a legitimate ground for certain forms of prohibition (see Feinberg 1986, 320ff.). Bearing all this in mind, it has to be concluded that the legitimate public policy on the enhancement use of abusable forms of methylphenidate (instant release) and amphetamines (both instant and extended release) by healthy adults in the general populace is prohibition of production and sale.

4.6 Conclusion

The use of Adderall (mixed amphetamine salts) and Ritalin (methylphenidate) by healthy adults for enhancement of cognitive function has to be dissociated from both therapeutic and recreational uses of these drugs. Also, regulation of their enhancement use has to be made while taking into account relevant differences in the danger profile. On the one hand, extended-release formulas of methylphenidate (e.g., Ritalin-SR) could be regulated permissively, since they cannot be recovered by readily applicable means in a quantity liable to abuse, and apparently do not give rise to a public health and social problem. The taxation approach (see Dubljević 2012a and Chap. 1 in this book) to regulation of cognitive enhancement drugs is a good starting point for such a moderately liberal public policy that avoids the pitfalls of both laissez-faire and overly harsh prohibitive policies (for discussions on these pitfalls see Capps 2011; Dubljević 2013, and Chap. 2 in this book). However, not all models of regulation within the broad taxation approach would be both appropriate and legitimate. Only the Economic Disincentives Model (Dubljević 2012b, see also Chap. 1 in this book) explicitly envisions all the measures required by the UN Convention of 1971, which make it the most legitimate public policy on extended release formulas of methylphenidate for cognitive enhancement use by healthy adults.

On the other hand, the sale of instant-release formulas of methylphenidate (e.g., Ritalin) to healthy adults, along with all compounds containing amphetamine (e.g., Adderall) or its precursors that would produce amphetamine via normal metabolism (e.g., Captagon), would need to be prohibited. Although these substances might provide significant benefits if used responsibly, the danger of abuse and especially the threats of addiction, increased aggression, and erratic and violent behavior make their use a potential danger to others. However, the use and possession of small quantities of these substances without a prescription should be treated as a misdemeanor and punishable only by a fine, whereas unauthorized production and sale could be legitimately criminalized and treated as a felony with appropriate sanctions.[20]

Finally, it should be noted that the arguments presented here cannot resolve the issue, since the relevant questions range from neurochemistry via moral philosophy to international law, and it is very hard for one or a few persons to cover them all to the full extent that they deserve. This limited "case analysis" should be understood as an invitation to experts in various spheres, citizens, and their political representatives to participate in an open discussion in the public forum, in which reliable data on consumption and demand, known effects, and relevant social implications and normative frameworks are presented and analyzed.

References

Auvin, S., et al. 2018. Systematic review of the screening, diagnosis, and management of ADHD in children with epilepsy. Consensus paper of the Task Force on Comorbidities of the ILAE Pediatric Commission. *Epilepsia* 59 (10): 1867–1880.

Barkley, R.A., M. Fischer, L. Smallish, and K. Fletcher. 2003. Does the treatment of attention-deficit/hyperactivity disorder with stimulants contribute to drug use/abuse? A 13-year prospective study. *Pediatrics* 111 (1): 97–109.

Beauchamp, T.L., and J.F. Childress. 2009. *Principles of biomedical ethics* (6th ed.). New York, NY: Oxford University Press.

Bigelow, B.C. 2006. *UXL encyclopedia of drugs and addictive substances*. Detroit: UXL/Thompson Gale.

Blank, R. 2010. Globalization: Pluralist concerns and contexts. In *Scientific and philosophical perspectives in neuroethics*, ed. J. Giordano and B. Gordijn, 321–342. Cambridge, UK, New York: Cambridge University Press.

BMA. 2007. *Boosting your brainpower: Ethical aspects of cognitive enhancements. A discussion paper from the British Medical Association [BMA]*. http://www.bma.org.uk/ap.nsf/AttachmentsByTitle/PDFCognitiveEnhancement2007. Accessed 23 Mar 2012.

Breggin, P.R. 2001. *Talking back to Ritalin: What doctors aren't telling you about stimulants and ADHD*. Cambridge, MA: Da Capo Press.

Capps, B. 2011. Libertarianism, legitimation, and the problems of regulating cognition-enhancing drugs. *Neuroethics* 4 (2): 119–128.

Carter, A., and W. Hall. 2012. *Addiction neuroethics: The promises and perils of neuroscience research on addiction*. Cambridge: Cambridge University Press.

Carter, A., W. Hall, and J. Illes (eds.). 2012. *Addiction neuroethics: The ethics of addiction neuroscience research and treatment*. London: Academic Press.

Coenen, C., M. Schuijff, and M. Smits. 2011. The politics of human enhancement and the European Union. In *Enhancing human capacities*, 676–693. Oxford, UK: Blackwell.

Cortese, S. 2018. Are the effects of methylphenidate uncertain? *Irish Journal of Psychological Medicine* 35: 163–167.

[20]Increased awareness about harmfulness of amphetamines and instant release forms of methylphenidate has led to a shift in preference among some student populations and even medical practitioners to newer, atypical stimulant drugs such as modafinil (see Dubljević and Ryan 2015; Dubljević 2016). The fact that newer drugs with stimulant-like properties were not rated at all in any of the MCHDS studies (see Table 4.2) significantly limits the debate on evidence based policy options for this class of drugs. Therefore, the discussion here is limited to 'classical stimulants' and more work needs to be done to apply such discussion to newer classes of drugs.

De Greiff, P. (ed.). 1999. *Drugs and the limits of liberalism*. New York: Cornell University Press.
De Marneffe, P. 2005. Against drug legalization. In *The legalization of drugs: For and against*, ed. D. Husak and P. De Marneffe, 109–198. New York: Cambridge University.
DeSantis, A.D., E.M. Webb, and S.M. Noar. 2008. Illicit use of prescription ADHD medications on a college campus: A multimethodological approach. *Journal of American College Health* 57 (3): 315–324.
Dubljević, V. 2012a. Principles of justice as the basis for public policy on psycho-pharmacological cognitive enhancement. *Law, Innovation and Technology* 4 (1): 67–83.
Dubljević, V. 2012b. Toward a legitimate public policy on cognition-enhancement drugs. *American Journal of Bioethics – Neuroscience* 3 (3): 29–33.
Dubljević, V. 2013. Prohibition or coffee-shops: Regulation of amphetamine and methylphenidate for enhancement use by healthy adults. *American Journal of Bioethics* 13 (7): 23–33. https://doi.org/10.1080/15265161.2013.794875.
Dubljević, V. 2016. Enhancement with modafinil: Benefiting or harming the society? In *Cognitive enhancement: Ethical and policy implications in international perspectives*, ed. F. Jotterand and V. Dubljević. New York: Oxford University Press.
Dubljević, V., and C.J. Ryan. 2015. Cognitive enhancement with methylphenidate and modafinil: Conceptual advances and societal implications. *Neuroscience and Neuroeconomics* 4: 25–33.
Duke, S.B., and A.C. Gross. 1993. *America's longest war: Rethinking our tragic crusade against drugs*. New York: G.P. Putnam's Sons.
Euromonitor. 2011. Tobacco in Norway. *Euromonitor International Market Research Report*. Available at: http://www.euromonitor.com/tobacco-in-norway/report.
EMCDDA. 2010. *Problem amphetamine and methamphetamine use in Europe*, Selected issue, European Monitoring Centre for Drugs and Drug Addiction, Lisbon. http://www.emcdda.europa.eu/attachements.cfm/att_120112_EN_EMCDDA_SI10_Amphetamines.pdf. Accessed 11 Apr 2012.
EMCDDA-Europol. 2011. *Amphetamine: A European Union perspective in the global context*. Joint publication of the European Monitoring Centre for Drugs and Drug Addiction and Europol. Luxembourg: Publications Office of the European Union.
Feinberg, J. 1986. *Harm to self (The moral limits of the criminal law Vol. 3)*. New York: Oxford University Press.
Fitzgerald, T.D. 2009. *White prescriptions? The dangerous social potential for Ritalin and other psychotropic drugs to harm black boys*. Boulder: Paradigm Publishers.
Freeman, S. 1999. Liberalism, inalienability and rights of drug use. In *Drugs and the limits of liberalism*, ed. P. De Greiff, 110–130. New York: Cornell University Press.
Freye, E. 2009. *Pharmacology and abuse of cocaine, amphetamines, ecstasy and related designer drugs*. Dordrecht: Springer.
Glannon, W. 2008. Psychopharmacological enhancement. *Neuroethics* 1 (1): 45–54.
Greely, H. 2011. Of nails and hammers: Human biological enhancement and U.S. policy tools. In *Enhancing human capacities*, 653–675. Oxford, UK: Blackwell.
Greely, H., B. Sahakian, J. Harris, R.C. Kessler, M. Gazzaniga, P. Campbell, and M.J. Farah. 2008. Towards responsible use of cognitive-enhancing drugs by the healthy. *Nature* 456 (7223): 702–705.
Husak, D. 2005. For drug legalization. In *The legalization of drugs: For and against*, ed. D. Husak and P. De Marneffe, 3–108. New York: Cambridge University Press.
Husak, D. 2007. Why we should decriminalize drugs. In *Ethics in practice*, 3rd ed, ed. H. La Follette, 334–344. Oxford: Blackwell Publishing.
Hyman, S. 2011. The neurobiology of addiction: Implications for the voluntary control of behaviour. In *The Oxford handbook of neuroethics*, 203–218. Oxford, UK: Oxford University Press.
Ilieva, I., J. Boland, and M.J. Farah. 2013. Objective and subjective cognitive enhancing effects of mixed amphetamine salts in healthy people. *Neuropharmacology* 64 (1): 496–505.
Iversen, L. 2008. *Speed, ecstasy, ritalin: The science of amphetamines*. Oxford: Oxford University Press.

Kass, L. 2003. *Beyond therapy: Biotechnology and the pursuit of happiness; a report*. New York, NY: Dana Press.
Kociancic, T., M.D. Reed, and R.L. Findling. 2004. Evaluation of risks associated with short- and long-term psychostimulant therapy for treatment of ADHD in children. *Expert Opinion on Drug Safety* 3 (2): 93–100.
Lee, N.K., L. Jenner, A. Harney, and J. Cameron. 2018. Pharmacotherapy for amphetamine dependence: A systematic review. *Drug and Alcohol Dependence* 191: 309–337.
Liang, E.F., S.Z. Lim, W.W. Tam, C.S. Ho, M.W. Zhang, R.S. Mcintyre, and R.C. Ho. 2018. The effect of methylphenidate and atomoxetine on heart rate and systolic blood pressure in young people and adults with Attention-Deficit Hyperactivity Disorder (ADHD): Systematic review, meta-analysis, and meta-regression. *International Journal of Environmental Research and Public Health* 15 (8): 1789.
Lieb, K. 2010. *Hirndoping: Warum wir nicht alles schlucken sollten*. Mannheim: Artemis & Winkler.
Liu, H., W. Feng, and D. Zhang. 2018. Association of ADHD medications with the risk of cardiovascular diseases: A meta-analysis. *European Child & Adolescent Psychiatry*.
Maher, B. 2008. Poll results: Look who's doping. *Nature* 452 (7188): 674–675.
Merkel, R., G. Boer, J. Fegert, T. Galert, D. Hartmann, B. Nuttin, and B. Rosahl. 2007. *Intervening in the brain: Changing psyche and society*. Berlin/Heidelberg: Springer.
Miller, R.L. 2002. *Encyclopedia of addictive drugs*. London: Greenwood Press.
Nutt, D., L.A. King, W. Saulsbury, and C. Blakemore. 2007. Development of a rational scale to assess the harm of drugs of potential misuse. *The Lancet* 369 (9566): 1047–1053.
Racine, E. 2010. *Pragmatic neuroethics: Improving treatment and understanding of the mind-brain*. Cambridge, MA: MIT Press.
Ragan, C.I., I. Bard, and I. Singh. 2012. What should we do about student use of cognitive enhancers? An analysis of current evidence. *Neuropharmacology* 64: 588–595.
Sandberg, A., W. Sinnott-Armstrong, and J. Savulescu. 2011. Cognitive enhancements in court. In *The Oxford handbook of neuroethics*, ed. J. Illes and B. Sahakian, 273–284. Oxford: Oxford University Press.
Singleton, J., L. Degenhardt, W. Hall, and T. Zabransky. 2009. Mortality among amphetamine users: A systematic review of cohort studies. *Drug and Alcohol Dependence* 105: 1–8.
Staatsblad. 2002. Opium act. Cannabis Bureau archive. Available at: http://www.cannabisbureau.nl/en/doc/pdf/Dutch%20OpiumAct30556.pdf.
Tarrant, N., M. Roy, S. Deb, S. Odedra, A. Retzer, and A. Roy. 2018. The effectiveness of methylphenidate in the management of Attention Deficit Hyperactivity Disorder (ADHD) in people with intellectual disabilities: A systematic review. *Research in Developmental Disabilities* 83: 217–232.
United Nations Office on Drugs and Crime. 2007. *Preventing amphetamine-type stimulant use among young people*. New York, NY: United Nations Publication.
United Nations [UN]. 1971. *Convention on Psychotropic Substances*. www.unodc.org/pdf/convention_1971_en.pdf. Accessed 3 Apr 2013.
Walitza, S., B. Werner, M. Romanos, A. Warnke, M. Gerlach, and H. Stopper. 2007. Does methylphenidate cause a cytogenetic effect in children with ADHD? *Environmental Health Perspectives* 115 (6): 936–940.
Wilson, J.Q. 2007. Against the legalization of drugs. In *Ethics in practice*, 3rd ed, ed. H. La Follette, 330–334. Oxford: Blackwell Publishing.
Yan, L., J. Zhang, Y. Yuan, and S. Cortese. 2018. Effects of neurofeedback versus methylphenidate for the treatment of attention-deficit/hyperactivity disorder protocol for a systematic review and meta-analysis of head-to-head trials. *Medicine* 97 (39).

Chapter 5
Legitimate Public Policies on Electro-magnetic Cognitive Enhancements

Drugs are not the only means of brain modulation. Indeed, medical devices have long since known to be able to modulate, and perhaps even improve, cognition. However, there has been a considerable amount of speculation regarding what kinds of medical devices might offer cognitive enhancement (CE) and what would be the accompanying ethical and regulatory challenges (see e.g., Maguire and McGee 1999; EGE 2005; Warwick 2008). Both utopian and dystopian claims about neuroimplants and brain-computer interfaces have been advanced, with little regard to real-world feasibility. However, relevant scientific and ethical reviews reported a potential for enhancement use of non-invasive brain stimulation technologies in healthy adults, which makes them a more realistic social issue. Transcranial magnetic stimulation (TMS) is frequently analyzed in the literature regarding ethical and regulatory aspects of CE (e.g. STOA 2009) and more than 60 academic articles report performance enhancements with TMS in healthy human subjects (Luber and Lisanby 2013).

Transcranial direct current stimulation (tDCS) has also gained prominence because of CE possibilities (Dockery et al. 2009), and this was also recognized by the media (see e.g., Adee 2012; Dubljević et al. 2014). A special issue of the neuroscience journal *Neuroimage* provided a review of studies confirming enhancement properties of tDCS in a multitude of cognitive tasks on healthy adults (Clark and Parasuraman 2013). Due to low costs and ease of production and use, tDCS devices can be widespread, and this has caused an upsurge in calls for more regulation (see e.g., Fitz and Reiner 2013; Anonymous 2013; Bikson et al. 2013; Maslen et al. 2013a, b).

Therefore, enhancement uses of both tDCS and TMS need to be taken seriously as an emerging social phenomenon, and their ethical, social and legal implications require careful philosophical consideration, public discussion and adequate regulation. However, regulatory options are limited by the immediacy of the social problem, efficacy of proposed solutions, and plausible types of social penetration. tDCS and

This chapter draws and expands on my previous work which has been published as: Dubljević (2015).

V. Dubljević, *Neuroethics, Justice and Autonomy: Public Reason in the Cognitive Enhancement Debate*, The International Library of Ethics, Law and Technology 19, https://doi.org/10.1007/978-3-030-13643-7_5

TMS (see Pascual-Leone et al. 2011, for a review), although fairly established as investigatory tools in neuroscience, have very different profiles of social penetration—a fact that needs to constrain ethical debates and analysis of policy options. For instance, the high costs and necessary technical knowledge (Simpson et al. 2009), make the issue of "wide-spread" use of TMS as a product for CE only hypothetical. However, the technique could have substantial social penetration as an off-label therapeutic and enhancement service. Furthermore, in cases where different modalities of social penetration are possible, reasonable policy options need to take all of them into account.

Media hype and misinformation are also important considerations that need to be addressed. Moreover, the lack of existing regulation (for non-therapeutic uses) to build upon is a problem for assessing the intended and unintended outcomes of policy. For example, tDCS has caused excitement in the lay public and academia as a "portable, painless, inexpensive and safe" (Cohen Kadosh et al. 2012) enhancement device. TDCS not only has the potential to become the tool for wide-spread CE, and create social problems meriting social regulation, the media hype surrounding it and emergence of commercial applications (Anonymous 2013) make the issue urgent. Unlike CE drugs where the regulatory environment is specified by international law (see, e.g., UN 1971), regulatory environment for enhancement devices is much less clear. As producers are quick to capitalize on the lack of regulation, there might be no time for fine-tuning policy: a comprehensive regulatory framework that could facilitate responsible use needs to be established urgently. After a legitimate policy is enacted and the legal default ("everything is permissible unless regulated") is replaced, it can be superseded with more efficient regulatory responses. The key issue is to generate sufficient information about the challenges at hand.

5.1 The Case of Transcranial Direct Current Stimulation (tDCS)

The history of tDCS is long and complicated: its development is marked by recurring fits and starts (see Fregni 2005). Successful applications were several times followed by a short-lived upsurge of attention in the medical community. However, with the advent of psychopharmacological interventions, the mainstream interest in tDCS and other "electrotherapies" waned. The advances in localization and standardization of study parameters late in the 20th century led to a renewed interest in tDCS (see Brunoni et al. 2011a). In fact, a range of factors contributed to a major resurgence of tDCS as an investigatory, therapeutic and, finally, enhancement technique (see Dubljević et al. 2014).

To start with, standardization allowed for replicability of results, and the possibility of combining this neurostimulation technique with guidance by neuroimaging devices made tDCS an effective tool of neuroscientific investigation (Clark et al. 2012). Also, the efficacy of some pharmacological therapies has reached its limit—for instance, certain cases of major depressive disorder resist or do not react well to pharmacological treatment—and this lead to a search for supplementary evidence-based therapeutic approaches (see Priori 2003). Finally, documented enhancement effects, first in studies on implicit classification learning (Kincses et al. 2003) and visuo-motor learning (Antal et al. 2004) led to a rising tide of interest, first in the scientific community, followed by military funding agencies, media outlets and the general public (see Adee 2012; Dubljević et al. 2014).

The replicability of CE effects in animals and healthy adults (Dockery et al. 2009, 2011; Hamilton et al. 2011; Clark and Parasuraman 2013), virtually guarantees that tDCS will not fall into oblivion again. Most notably, the findings of therapeutic effects in depression (Fregni 2005) and enhancement in mathematical ability (Cohen Kadosh et al. 2010) deservedly generate interest. Furthermore, the fact that these effects are long lasting and achieved with a technique which is perceived as non-invasive, cheap, easy to administer, and painless could influence the uptake in social penetration. Military agencies such as Defense Advanced Research Projects Agency (DARPA) in the U.S. have fueled the interest by offering extensive funding (see e.g., Clark et al. 2012), whereas the media have enthusiastically suggested to the public unproven (see e.g., Dubljević et al. 2014).[1]

Apparently, the social penetration of tDCS as a therapeutic, investigative and above all enhancement device can be seen as rapidly increasing (Dubljević et al. 2014). However, if a certain device like tDCS could be used as enhancement and there is a possibility of wide-spread use, this does not necessarily mean that it is morally problematic. Only if this use generates significant negative effects is a regulatory response of the state justified.

In order to understand the complexity of tDCS, this section of the chapter will shortly review the mechanism of action, uses and adverse effects. tDCS is a non-invasive technique which uses low-intensity direct current to cortical areas in order to facilitate or inhibit spontaneous neuronal activity. It is primarily used as an investigative and therapeutic tool in the context of pain, depression, as well as neuro-rehabilitation and learning enhancement after stroke (see Fitz and Reiner 2013). Unlike stimulant drugs, which follow a posology with predictable effects, tDCS is a device that can be easily built at home and used repeatedly on different cortical locations and in various stimulation modalities (Fitz and Reiner 2013).

The establishment of safety protocols has been crucial for the re-invigoration of scientific interest in the technique (Priori 2003). Articles noting that use of tDCS is safe and effective (e.g., Poreisz et al. 2007) are referring specifically to established procedures in laboratory settings. However, in early investigative uses in the mid-twentieth century, tDCS was reported to cause serious adverse effects, such as temporary respiratory paralysis (Brunoni et al. 2011a). Several other detrimental effects are regularly caused even with strict adherence to safety protocols. Cases reported in the literature include tDCS induced dermatitis (Riedel et al. 2011), hypomanic episodes in depressed (Arul-Anandam et al. 2010) and bi-polar patients

---

[1]A notable example is: "Schoolchildren who struggle to grasp mathematics could benefit from having their brains roused with electricity"—see Sample (2010).

(Galvez et al. 2011), and even manic psychosis (Brunoni et al. 2011b) when combined with anti-depressant medication.

Furthermore, publication focus and bias might contribute to misunderstanding of tDCS effects. Academic articles for the most part paint an overly enthusiastic picture of tDCS (Dubljević et al. 2014) because potentially detrimental effects are not measured and hence not reported, although tDCS enhances certain functions while at the same time inhibiting others (Iuculcano and Cohen Kadosh 2013). A meta-analysis of tDCS studies in motor and cognitive domains (Jacobson et al. 2012) reported that during stimulation one of the electrodes used (anodal electrode) has the effect of enhancing cortical excitability whereas the other (cathodal electrode) actually diminishes it.[2] But what is known about how tDCS induces changes, and what are the safety protocols for use?

According to Stagg and Nitsche (2011), safe use of tDCS entails duration between 10 and 20 min and using two surface-conductive electrodes (anodal and cathodal) sized between 25 and 35 cm^2 (which translates to roughly 4–5 $in.^2$). The size is important because the electrical current is distributed along the electrode surface: the electricity enters through the anode, passes through brain tissue and exits through the cathode. Accordingly, the smaller the surface, the higher the current density in the stimulated region even with the same current intensity (usually 1–2 mA).

As for how exactly tDCS works, Brunoni et al. (2011a) report that the mechanisms of action likely involve different synaptic and non-synaptic effects on neurons and on non-neuronal (e.g., glial) cells and tissues within the brain. Long-lasting effects appear to depend on protein synthesis, beyond a mere phenomenon of electric transmission. Anodal stimulation seems to increase intra-neuronal levels of calcium and neurotransmitter-receptor dependent gene expression (see Stagg and Nitsche 2011). However, the primary mechanism of action is the polarization of resting membrane potential (see Brunoni et al. 2011a), with after-effects lasting for up to one hour. The synaptic micro-environment is also modified by tDCS, which affects excitatory (such as dopamine) and inhibitory neurotransmitters (such as gamma-Aminobutyric acid or GABA). Since most neurotransmitters and neuro-receptors have electrical properties, and tDCS causes a constant electric field which displaces all electrically charged molecules, it might induce additional prolonged neurochemical changes (see Brunoni et al. 2011a).

A device with unclear mechanisms of action and potential detrimental effects need not create problems in all instances. Additional constraints for social penetration (i.e., preventing wide-spread use) might be high costs and low availability. However, tDCS promises extreme availability, since it is currently advertised online as an enhancement product, service, and as an easy to make Do-It-Yourself (DYI) gadget (see Dubljević et al. 2014). Due to one-sided, overly enthusiastic portrayal of tDCS in the media, the risks are likely to be poorly understood by the general public (Dubljević et al. 2014). Accordingly, any prospective regulatory framework would

---

[2]The inhibitory effect of cathodal stimulation seems to be consistent in motor studies; in the cognitive category, memory and executive function are affected, whereas the evidence on effects on language ability is not conclusive (see Jacobson et al. 2012).

need to address these three types of social availability of tDCS, particularly in context of use on developing brains (e.g., children).

The apparent effectiveness, re-usability and low costs of tDCS suggest a high penetration rate of this technology and tremendous impact in the clinical context and in competitive social milieus (Business, Education, Military, etc.). Therefore, it is safe to assert that a discussion on regulation of tDCS is timely and not premature. There have been some academic papers on the issue (see e.g., Fitz and Reiner 2013; Anonymous 2013; Bikson et al. 2013; Maslen et al. 2013a, b), however, the focus of these articles was on the do-it-yourself enhancements, and in response to the tDCS neurogaming product, and not on other products that could be used in the tDCS modality or the regulation of tDCS as a service. Since the latter two are on the rise, this presents an important gap.

Despite recommendations and calls for extending existing legislation for medical devices to tDCS as products (see e.g. Fitz and Reiner 2013; Maslen et al. 2013a, b, 2014), no concrete proposals for regulatory framework for all modalities (product, service, and DIY gadget) of tDCS use by healthy adults have been advanced. This issue merits attention from the academic community and policy makers, particularly given the rapid evolution of neuromodulation techniques. Two approaches that were discussed in the context of CE drugs could potentially be extended to tDCS: the "gatekeeper" model (BMA 2007) and the Economic Disincentives Model (EDM).[3]

It should be remembered that the "gatekeeper" model advises relying on health professionals to act as "gatekeepers" of medical technologies that could be used for CE (BMA 2007). Indeed, it could be argued that health professionals should bear in mind all available information about tDCS when making the decision whether or not to allow it to "enhancement seekers". After all, the American Neurological Academy has issued an influential set of guidelines which concluded that medical doctors have the right to decide whether to prescribe drugs as enhancement or not based on their expertise and good medical practice (Larriviere et al. 2009). Perhaps this advice can be extended to medical devices as well, with or without additional guidance and licensing from professional bodies.

However, there are several specific problems with such a "gatekeeper" approach. If this model was used for "prescribing" tDCS as a product, once the device has been purchased it is beyond the control of the medical professional (or anyone else for that matter). The problem of untrained use on any number of individuals remains, as unlike prescription drugs, tDCS is reusable indefinitely. If tDCS in fact can be used responsibly, by taking into account safety settings and procedures, the "gatekeeper" model actually cannot be used for tDCS as a product, only for regulation of commercial uses of tDCS as a medical *service*. In case that this is the only legally available solution to gain access to tDCS, this might lead to further problems. In Chaps. 1 and 2, I have argued that medical doctors have the expertise to diagnose

[3] As the reader might recall, the "gatekeeper" model has already been rejected in Chaps. 1 and 2. However, the pertinent issue here is that electro-magnetic enhancement devices such as tDCS can be used in different modalities (i.e., not just as a product), unlike psycho-pharmacological enhancements.

illnesses and prescribe therapy, whereas every citizen should have the right to decide for her/him-self whether to use enhancements or not (see also Dubljević 2012a, b, 2013). If citizens view laws and regulations as overly constraining and unnecessary, they are highly unlikely to obey them. Since tDCS is also available as a product or Do-It-Yourself gadget it is very likely that a number of people would take advantage of such availability to circumvent the costly "gatekeeper".

This does not lead to a conclusion that the approach is pointless—it only means that the "gatekeeper" model for the tDCS service would have to be amended by additional regulation. The direction (precise populations being regulated) and severity of sanctions connected to such additional regulation constrains the effectiveness of the model. Now, the gatekeeper model might be bolstered by sterner and more prohibitive regulation of tDCS as a product by the state. Perhaps possession and use of tDCS devices can be restricted to health professionals only, and any non-compliance sanctioned with criminal prosecution. Apart from the questionable legitimacy of such a response, two additional socially undesirable consequences would be produced. The first is creating a black market for tDCS, and the second is restricting access to tDCS to only a subset of population that could afford the high fees of medical practitioners. In short, that would make the "gatekeeper" model both ineffective and unjust, since the issues of paternalism and the accumulation of the power to distribute enhancements in the hands of health professionals make justification of this approach to all citizens very hard.

A different move could be to offer more permissive additional regulation. The "gatekeeper" model could be sufficient to license the individuals that would have the right to apply tDCS to others, but the issue that needs to be tackled is untrained use on oneself, which creates the problem in the first place. The "gatekeeper" model might be bolstered with stern regulation of unlicensed application of tDCS to others (especially children and adolescents) and relatively permissive regulations toward applying tDCS to oneself. Indeed, if the analogy with other technological advancements is taken into consideration, even potentially dangerous technologies have been rendered relatively safe by licensing of users with few restrictions apart from age and health. For instance, a piloting license can be gained by every adult citizen, providing that adequate training and health are proven. On a more permissive range, driver's licenses are available in some countries from the age of 16, and are again limited with certain medical conditions (e.g., sight impairments).

Since tDCS appears to be safe and effective in laboratory settings, it could be plausible to assume that it might be safe and effective in any environment with sufficiently trained users. The limits for tDCS use might not need to be as harsh as those for the piloting license, but should not be as permissive as in the case of driving licenses. Since it was noted that tDCS might cause long-term detrimental changes in developing brains, a reasonable precaution could be to initially set the age requirement at 25, and to amend the age requirement if the consensus in the scientific community about safety at earlier ages is reached.

A more permissive approach to tDCS would mean extending the mandate of regulatory agencies to non-therapeutic uses of stimulation devices as proposed by some authors (e.g., Maslen et al. 2013a, b). However, since the problem identified was untrained use, the mandate would need to be further extended to enforcing the necessary competence of would-be enhancement seekers in order to ensure safe and responsible use. The Economic Disincentives Model (EDM), which (the reader should remember) would introduce a licensing procedure and impose taxes, fees and requirements of additional insurance, could be extended from stimulant drugs to stimulation devices and might provide a solution to the problems identified above.

If a designated government agency offered a licensing procedure for companies to market tDCS for healthy adults, all citizens of age would be able to legally purchase tDCS devices. Since EDM also envisions that taxes, fees and requirements of additional insurance would be imposed, it could be questioned if the financial and regulatory burdens thus created would be the best solution. However, recall that there is a lack of information on effects and mechanisms of action of tDCS, and apart from limiting the availability of tDCS to minors, the regulatory framework needs to generate the information on harmful effects as soon as possible.

By envisioning the additional licensing procedure for users, EDM has the potential to provide this information in the shortest amount of time. Since in order to be able to purchase, possess and use a tDCS device, citizens would have to pay fees for a course about safety pre- cautions, proper operation, known effects and side effects of tDCS, and pass an exam as proof of knowledge, most detrimental effects could be avoided.[4] Furthermore, the additional medical insurance would guarantee that any adverse effects generated by tDCS use would not drain public funds. Finally, obligatory annual medical tests for obtaining and renewing a license to use a tDCS device would quickly generate the information needed for fine-tuning the policy.

Thus, there are good reasons to assert that EDM has the advantage of quickly, cost-effectively and objectively generating data for post-market monitoring. However, it should be noted that users would have the right to opt out from providing their data if they have privacy concerns. The requirement of EDM to regulate the prices of tDCS devices might turn out to be unnecessary. Namely, in the original formulation (see Chap. 1), EDM proposes regulating the prices by including the standard costs of production and distribution, limiting the profit margin and imposing an additional tax. Now, if the data from tDCS users points to the conclusion that trained use of tDCS is reasonably safe even outside controlled laboratory settings, these requirements, along with further taxing of the companies could be relaxed. However, they would need to be initially enforced. Thus, the considerable regulatory burdens for enhancement seekers would limit the social penetration until the issue of long-term physiological effects of tDCS has been settled by reliable data. Since tDCS would in principle be available to all, while regulatory burdens are discouraging its use and wide-spread social penetration, this should offset any concerns about fairness.

The discussion so far leaves the issue of non-commercial uses of tDCS unresolved. Technically, tDCS as a do-it-yourself gadget defies almost all efforts to regulate the

---

[4]The question of liability for harmful effects after the training has been provided is potentially salient, but easily answered in this context. Just as driving instructors are not liable for the traffic accidents made by their trainees after they pass the exams and receive their licenses, so tDCS trainers should not be held responsible for adverse effects caused by "trained and certified" individuals.

technology. However, having a reasonable legal alternative is enough in most cases to promote regulated use of safe products. Furthermore, apart from using tDCS on others, even unlicensed self-uses could be criminalized and a moratorium on Direct-to-Consumer marketing of tDCS could be enforced. However, such harsh measures should be avoided, unless there is proof that unlicensed home uses of tDCS pose a considerable danger.

5.2 The Case of Transcranial Magnetic Stimulation (TMS)

Does TMS need to be urgently regulated as well? This question is not so easy to answer. On the one hand, the social penetration of TMS as a product is likely to remain low since the costs are very high (see Simpson et al. 2009; Sutherland 2013). On the other hand, the use of TMS as a service might have a moderate social penetration, especially since enhancement effects have been reported in more than 60 studies (Luber and Lisanby 2013). But what exactly is TMS?

TMS is a technique that uses a device which stores a strong electrical charge in capacitors (Fitzgerald and Daskalakis 2013). Periodic discharge of the stored electrical energy from the capacitors passes a strong electrical current through a magnetic coil. The coil is placed over the specific position on the cranium of the subject (depending on the cortical location to be stimulated). The magnetic field thus generated passes through the cranium and induces a weak electric current inside the surface areas of the brain causing hyper- or hypo-polarization of the neurons in the affected brain tissue (Schermer 2014). This simple and very effective principle of non-invasive brain stimulation has the advantage over electrical forms of stimulation in that the magnetic field passes easily and virtually without any resistance through the skin and cranium, and generates the current within the brain structure itself (see Richter 2013).

TMS has been evaluated by the scientific and medical community in an impressive number of research studies and clinical applications (Rossi et al. 2009). Although it has been widely used in a plethora of scientific and clinical studies (see Boniface and Ziemann 2003) and research suggests that it can help people with post-traumatic stress disorder, bipolar disorder and Parkinson's disease, to date, TMS has only been approved to treat major depressive disorder to date (Sutherland 2013). The fact that the danger profile of TMS could be considered moderate, due to the risk of seizures, the most serious TMS-related acute adverse effect (Rossi et al. 2009), might have contributed to the lack of approval for the conditions mentioned above, even though it seems to be effective and might also be beneficial in a variety of other conditions, including schizophrenia, anorexia, Alzheimer's disease, autism and cerebral palsy (Sutherland 2013).

Furthermore, TMS has different effects and side-effects depending on: (1) the intensity of the stimulation, (2) modality of stimulation, (3) the duration of stimulation, (4) the intervals between stimulation sessions, and (5) the site being stimulated. Each of these parameters is shortly reviewed in the following sub-sections.

5.2.1 *Intensity*

In case of any form of TMS, the time-varying magnetic field is generated by a short high-current electric pulse (4–20 kA) with high voltages (400–3000 V) sent through the stimulation coil, with the resulting short term magnetic field (a few milliseconds) and peak strengths of 1–10 T (Richter 2013). When the coil is placed above the skull, the magnetic field will cause weak electrical current to flow in nearby secondary conducting material, such as neurons, and if this current is of sufficient strength, it will produce depolarization or hyperpolarization of the conducting neural tissue located just under the coil (Fitzgerald and Daskalakis 2013).

5.2.2 *Modality*

The modality of TMS is influenced by the choice of pulse types (single pulse or repetitive stimulation of low or high frequency), coil shape (circular, figure of eight, double cone or H-shaped), the choice of technique (paired pulse, theta burst, etc.). Single pulse TMS has been successfully used even with older mono-phasic stimulation devices (which had a delay of 3 s between charges) to disrupt neural activity and to distinguish between spinal and cortical effects in the nervous system (see Rothwell 2003). Both single pulse stimulation, which mostly lead to an immediate reaction (e.g. twitching of muscles) and repetitive TMS (rTMS), which can facilitate longer lasting changes to neuronal behavior (Richter 2013), can be achieved with the newer, bi-phasic stimulators, which have a considerably higher rate of recharging capacitors (Fitzgerald and Daskalakis 2013). The different frequencies of rTMS have diverse effects on neural tissue: research findings indicate that low frequency stimulation (below 5 Hz) is generally less likely to cause seizures and decreases neuronal excitability (Richter 2013), while high frequencies (greater than 5 Hz, but usually 10–20 Hz or more) generally excite the affected neuronal tissue with greater and longer lasting effects, but at the same time increase the danger of adverse effects (Rothwell 2003).

Beside the properties of the device that generates the charges, the shape of the stimulating coil also affects the magnetic field properties and, conversely, the effects of brain stimulation (Richter 2013). Initial TMS studies used circular or round coils (Fitzgerald and Daskalakis 2013). These coils generate a ring-like field, which is less focal, but depending on the size can penetrate deeper than other standard coils and stimulate structures of the brain beyond the immediate surface of the brain (see Rossi et al. 2009). Such larger and deeper fields may be preferred when the desired neuro-anatomic stimulation site is not precise, but encompasses a greater cortical area (Fitzgerald and Daskalakis 2013). More focal stimulation requires 'figure of eight' coils, which consist of two circular coils in a single plane, and this shape has the effect of adding the two induced fields at the intersection (Richter 2013), thereby providing better control over the site being stimulated (Fitzgerald and Daskalakis

2013). The ‘double cone’ coil is a variation of the figure of eight coils, as it is formed by two large circular wings. The wings are placed at an angle of 95°, which enables induction of a stronger field, which penetrates the surface and allows for stimulation of regions located deeper in the brain (see Rossi et al. 2009).

Apart from the more common types of coils, the need to produce more focal effects in the deeper structures of the brain during so-called deep TMS (see Bersani et al. 2013) has motivated the introduction of more specialized coil designs. Most notably, after rigorous testing at high intensities and frequencies of stimulation, the specialized ‘H-coil’ has been approved for use on human subjects in Europe (Rossi et al. 2009). By virtue of having multiple coil windings, the H-coil generates sufficient magnetic field strength at distances of 6 cm or 2–2.5 in. (Fitzgerald and Daskalakis 2013)—a major improvement in relation to most conventional coils, which have the effective distance of 1.5–2.5 cm or roughly 0.5–1 in. (Bersani et al. 2013).

The final issue in regard to modality of TMS use is the choice of technique. The most notable specialized techniques are paired pulse and theta burst stimulation (TBS). Paired pulse is an old and reliable technique often used in investigative applications of TMS, and can be used even with older monophasic stimulators. The technique involves the application of two stimuli separated by a varying inter-stimulus interval in order to establish a functional connection between two different brain sites (Rothwell 2003).

Theta burst stimulation (TBS) can only be used with repetitive TMS (rTMS) and hence newer, biphasic stimulators. This technique involves several high frequency bursts which are delivered in short intervals (Rioult-Pedotti and Donoghue 2003). The virtue of the TBS paradigm is that it is able to produce longer effects with shorter stimulation times: Richter (2013, p. 4) reports that 200 intervals of short, high frequency bursts (50 Hz), with average 3 pulses and two-second interval pause are commonly used. The pauses between the bursts mitigate the effects of heating and reduce the danger of seizures as long as safety protocol, which limits the total number of pulses to 600 and stimulation intensity to 60% of device output, is observed (Rossi et al. 2009).

5.2.3 *Duration of Stimulation*

Duration of TMS refers to two separate issues: duration of a TMS session or a duration of the whole TMS treatment. A stimulation session consists of several stimulation trains (or a number of bursts in case of TBS). The increase in knowledge and refinement of technology have changed the length of a typical TMS session: initially 10–20 stimulation trains per session were considered safe, but in this decade studies 75 or more trains are the norm (Fitzgerald and Daskalakis 2013), and the most common duration of a TMS session is 15–30 min (Richter 2013). Duration of the whole treatment has evolved as well, but the safety profile in this respect is less clear, as cumulative daily or weekly application of TMS still needs to be assessed in terms of safety for both patient populations and healthy adults (see Rossi et al.

2009). Initially, 1 or 2 weeks of stimulation have been considered safe, but in the last five years treatments of 6 weeks or more have been conducted (Fitzgerald and Daskalakis 2013). Since a greater number of sessions is necessary to prolong the effects of stimulation (Schermer 2014), this trend of increase in TMS duration is likely to continue.

5.2.4 Interval Between Repeated Sessions

The scientific and medical community has not established an upper limit for the interval between sessions. However, safety guidelines, which have been established at a consensus meeting of most influential academic researchers using TMS around the world, require at least a 15 min interval between rTMS sessions (see Rossi et al. 2009).

5.2.5 Brain-Site Being Stimulated

The cortical location being stimulated is of utmost importance to the effects of stimulation (similarly to the case of tDCS). Initially, TMS could stimulate only the surface of the brain at various locations, but with the advent of deep TMS the variability of precise stimulation sites in more basic structures of the brain has increased (see Bersani et al. 2013).

These five sources of considerable variability of application (as reviewed above) raise the issues of effectiveness of TMS as an enhancement tool, known side-effects and dangers of TMS and the extent of social penetration that can be expected. As mentioned above, more than 60 studies report CE effects of conventional TMS in perceptual discrimination, motor learning, visual search and tasks involving attention, memory and language in healthy human subjects (see Luber and Lisanby 2013), whereas 5 studies report reliable enhancement effects of cognitive capacities of deep TMS (Bersani et al. 2013).

Moreover, scientific reports are increasingly focused on possibilities of augmenting cognition of human operators of complex technological systems. For example, a review of TMS, which acknowledged funding and direction from the military, explored the ability of TMS to improve "the weakest link in the system" by "direct augmentation of human performance" (McKinley et al. 2012, p. 130). Military funding agencies such as DARPA are very interested in funding studies in order to expand technical capabilities and reduce the number of personnel required to process vast amounts of data (McKinley et al. 2012). As scientific studies and TMS applications are publicly available, it is more or less uncontroversial to assume that the sphere of business is likely to pick up this trend.

Additionally, even though 'deep TMS' is fairly recent, standard versions of TMS have been available much longer and have been tested and evaluated for enhancement effects. On the one hand, similarly to most CE drugs, TMS can provide "performance maintenance" effects (Ranisch et al. 2013) such as reducing the effects of sleep deprivation on working memory (Luber et al. 2013). On the other hand, TMS also offers genuine increases in performance in a range of cognitive capacities, from analogical reasoning (Boroojerdi et al. 2001), to savant-like abilities (Snyder et al. 2006; Snyder 2009).

However, the exact mechanisms by which TMS enhances cognition are unknown, and although TMS of frontal brain regions modulates the activity of a range of neurotransmitters (Keck et al. 2000), it was generally considered to disrupt neural activity as in "virtual lesion" studies (Boniface and Ziemann 2003). That is why at first the enhancement effects were seen as a sort of puzzle, leading to several hypotheses.[5]

Whatever the exact mechanisms might be, the available evidence points to the conclusion that TMS can produce enhancement effects, but only under supervision by trained professionals. Furthermore, the serious nature of potential side-effects further necessitates a controlled environment for any kind of TMS use, and especially for enhancement uses.[6] The need for professional oversight might change in the future, with the advent of robotized neuro-navigation TMS systems (see Richter 2013), but this is likely to increase the costs of an already very expensive technique.

It is fairly uncontroversial to assume that, due to high costs and necessary level of expertise required (see Simpson et al. 2009), enhancement with TMS might be available only as service and not as a product in the near future. Even TMS service is expensive: a single TMS therapeutic session costs 300 USD, and since total duration of a TMS treatment include 20–30 sessions, the total cost of the TMS service ranges between 6000 and 10,000 USD (Sutherland 2013). Obviously, the social penetration of TMS can be expected to be moderate, and at first enhancement with TMS would be available only to the more affluent members of the society.

Whatever the extent of social penetration may be (as technological advances may lower the prices and miniaturization might increase availability) the appropriate regulation of TMS has to be cautious, and sale of TMS as a product to untrained individuals or enthusiastic groups may have to be explicitly prohibited and backed by stern sanctions. Caution is advisable not only due to known side-effects (it is more or less uncontroversial to assume that with untrained users the side-effects are much more likely), but also due to the possibility of not yet reported adverse cognitive effects and cumulative long term effects. Namely, as was the case with tDCS, the scientific evidence points to the conclusion that TMS is capable of producing CE

[5]Luber and Lisanby (2013) have reviewed evidence on three different kinds of enhancement effects, along with current hypotheses for explaining the phenomena: (i) enhancement via nonspecific effects of TMS; (ii) enhancement via "addition by subtraction," and (iii) enhancement effects via direct TMS to specific task-related brain regions. Due to reasons of space, I will only note these hypotheses and point the reader to the original source. For the purposes of this discussion it is sufficient to note that cognitive enhancement effects are, in fact, empirically observable.

[6]Compared to tDCS, TMS is much more dangerous. The danger of causing seizures is present even in controlled laboratory settings, let alone in "at home" uses, which basically justifies a more restrictive approach.

for the specific tasks measured in the studies, but it is not known if the enhancement effects come at the price of reduction in other cognitive functions and tasks (see McKinley et al. 2012). Also, adverse effects of long term (months or years) repeated exposure to TMS cannot be excluded. This includes not only unknown complications due to TMS induced fields, but also other effects such as hearing impairments, since TMS coils produce noise levels during operation which may exceed 140 dB (see Rossi et al. 2009).

Current regulation of TMS explicitly assumes its use as a therapeutic and investigatory device, which leaves the enhancement uses unspecified and unregulated. There is an expert consensus on guidelines on the use of TMS in three broad classes, under the mandate of Internal Review Boards of research institutions or Ethics Committees of medical facilities (see Rossi et al. 2009).

However, all three classes (as defined by Rossi and colleagues) exclude enhancement uses of TMS. It could be assumed that such expert consensus on guidelines for use of TMS would be enough to adequately curtail any enhancement uses until more reliable information on risks and benefits of specific TMS stimulation paradigms and frequencies are obtained.

However, this assumption would be overly optimistic. Consider the example of neuroimaging: even though there is a consensus among leading experts that the use of functional magnetic resonance imaging outside of controlled laboratory and clinical settings is premature (see Illes and Sahakian 2011), that has not stopped companies and private individuals from using this expensive technology in market research (see http://www.nmsba.com/what-is-neuromarketing), for purposes of commercial lie detection (http://www.noliemri.com) and even as evidence in criminal court proceedings (see Patel et al. 2007).

The use of cognitive enhancers in the courtroom is a case in point. For example, there is evidence that "addition by subtraction" uses of TMS might provide increased resistance to false memories (see Snyder 2009). Given that veridical memory is of utmost importance in witness testimony, there have been calls for mandatory use or government incentives for the use of CE in criminal proceedings of capital cases (Sandberg et al. 2011). At any rate, impressive effects of TMS have been reported by the media (see e.g., Sutherland 2013), so it would be naïve to assume that expert guidelines will prevent individuals or groups from using TMS to attempt to "unlock" their "inner savant," or enhance certain cognitive capacities. Indeed, relying on professional guidelines only is far from being conducive to curbing enhancement practices—it merely opposes government involvement.[7] There is much to be said about caution toward government involvement in liberal democracies. However, government involvement in the case of TMS is already relatively high, at least in some countries.

Namely, military interest (and funding) for TMS research (see McKinley et al. 2012) will certainly pave the way to more affordable TMS applications, and with the

[7]The reliance on professional guidelines is problematic in two separate respects: 1. Such guidelines do not exert sufficient regulatory control, and 2. The lack of state involvement exacerbates the problem of lack of control.

"spill-over" effect these will make their way into the general population either by off-label uses or by enthusiastic pro-enhancement groups practicing "investigatory uses" on their own. Even though TMS is likely to become more affordable in the future, it is highly unlikely that it will be available as a "do-it-yourself" enhancement gadget the same way tDCS is. This leaves two issues for regulation: marketing TMS devices as a products and TMS sessions as a commercial service.

So what kind of regulation would be appropriate for TMS as a product or as a service? There have been some attempts to answer this question. For example, Blank (2013) contends that the risks TMS poses are moderate, enhancement potential unknown, and that an appropriate cautious policy would be prohibition until more data is available. I disagree with Blank's conclusion regarding the enhancement potential of TMS and, based on available data (see Luber and Lisanby 2013) conclude that TMS in certain modulations might offer enhancement effects. However, I agree with Blank that the sale of TMS devices would have to be prohibited to anyone apart from recognized research and medical institutions, and that any non-compliance should be sanctioned with criminal prosecution.

In the case of medical practitioners offering TMS as a service, the situation is more complicated. Notwithstanding the fact that all governments have a responsibility of ensuring safety and quality control standards as well as consumer protection and fair market practices, the case for prohibition of TMS as a service is not as strong. Blank's reliance on the current medical risk assessment system, which compares treatment risk with the expected benefit of reduced morbidity from successful treatment makes him overly risk averse if CE by TMS is offered by a trained professional. As Bostrom and Sandberg (2009) rightly note, cosmetic surgery has offered a precedent for a risk model where the client's autonomy overrides at least minor medical risks even when the procedure does not reduce or prevent morbidity. The key issues are that the service is offered by a trained professional, which reduces harmful effects, and that the autonomy of would-be enhancement seekers is not undermined. Even though the adverse effects of TMS can be considerable, to date, there is no data on any effects causing addiction, dependence, withdrawal and tolerance, or any long term debilitating effects on agency.

Furthermore, the existence of commercial tDCS applications (as product and service) on the market makes the issue of fairness of regulation of TMS untenable as an isolated case. Consider the regulatory approach whereby use is discouraged: if TMS was an isolated case, any discouragement policy would be unfair as it would make the technology inaccessible to the least well of, while more affluent citizens might still afford it. The fact that tDCS (a very affordable enhancement device) is already available offsets the concern for fairness in access in the case of TMS.[8]

[8]This assertion might need to be qualified. Namely, not everyone will see the trade-offs between enhancement effects and potential side-effects of tDCS and TMS in the same light. As I argue in Chap. 2, the lack of adequate information on long term effects and even short term benefits (the issue how laboratory findings of improvement in cognition relate to everyday performance is far from clear) complicates the comparison of evaluations of benefits, harms and positional (dis)advantages. Since they are at this point not fixed, the assumption regarding offsetting the concerns of fairness could be provisionally accepted, until disputed by further evidence.

Thus, TMS should be regulated in such a manner that tDCS and TMS regulation are complementary. Also, it has to be taken into account that compared to tDCS, TMS can be focused more precisely, but it is less easy to use, more expensive and riskier due to the very real danger of seizures (see Schermer 2014). Therefore, appropriate regulation should be designed to protect the interests of citizens, and guard against specific avoidable harms of TMS.

As I have argued that tDCS needs to be regulated urgently and analyzed two promising approaches (the "gatekeeper" model and EDM), the most logical step is to assess the appropriateness of similar regulation for TMS. Since TMS, like tDCS, might cause long-term detrimental changes in developing brains, a reasonable precaution would be to again set the age requirement for eligibility for CE with TMS at 25. Thus, even if TMS as a service is available via licensed trained professionals, the service should not be performed on any person below 25 years of age, and any non-compliance by service providers should be sanctioned with criminal prosecution and loss of license. On the other hand, due to respect for autonomy, under-age enhancement service seekers should only be fined, and not criminalized.

Therefore, it seems that a "gatekeeper" approach could be used for regulation of commercial uses of TMS as a service. Now, to consider if this approach would lead to further problems: As previously mentioned, it is highly unlikely that TMS would be readily available as a product or a Do-It-Yourself gadget, and so it is highly unlikely that people will try to circumvent the trained professionals acting as "gatekeepers". As for the issue of who could be a "gatekeeper", a licensing procedure could be defined, which should not be limited to health professionals. After all, neuroscientists might be more knowledgeable on the issue than, say, a general health practitioner. Thus, the licensing procedure for service providers (and the "competent service provider" would be a better designation of the model than the "gatekeeper"), should be open to individuals with specialized training in neurology or neuroscience. But would this "competent service provider" model produce the same socially undesirable consequences as was argued for the "gatekeeper" model in the case of tDCS?

More specifically, would a black market for TMS devices be created, and would only the rich be able to afford enhancement with TMS? The issue of affordability of TMS is important, in both instances. Namely, the TMS device is very costly and hard to produce, so the danger of the black market is limited, even though it cannot be entirely excluded. On the latter issue, recall that the military is interested in TMS and that military funded research is likely to lead to a decrease in the costs of TMS as a service. However, the most important issue is the relation between supply and demand—sufficient competition tends to decrease prices. Thus, the "service provider model", by virtue of being open to medical doctors and neuroscientists, is likely to create sufficient supply, and likely drive the TMS service prices down.

But what about the lack of information on effects and mechanisms of action of TMS? Apart from limiting the availability of TMS as a service to minors, how would this "service provider model" generate the necessary information without increasing already considerable costs?

The answer is relatively simple. Recall that EDM is already proposed for tDCS and has the potential to provide information on long term effects, as well as safety and efficacy, in a short amount of time. Although some requirements of EDM could create additional costs and thus should not be enforced separately on TMS, the capacity to generate information is very valuable for society. The most elegant solution is to use the licensing of tDCS users for this purpose. Namely, the required exam for the user license for tDCS could be expanded to include known long-term effects and side-effects of TMS as well, and TMS as a service should only be made available to such "neurostimulation enhancement device" license holders.[9] Therefore, there would be no need for a specific additional medical insurance for TMS, as one insurance policy would include both TMS and tDCS, and thus, again, this would guarantee that any adverse effects generated by TMS use would not drain public funds. Finally, the annual medical tests for obtaining and renewing a license would quickly generate the information needed for fine-tuning the policy.

Although there is the danger that the statistical data about unwanted effects and long term consequences of prolonged enhancement use might be confounded by use of multiple neurostimulation techniques (tDCS and TMS), this would only increase the time-span before relevant information is available, and is not a sufficient reason to abandon the model. Again, due to already high costs of TMS as a service, the requirement of EDM to regulate the prices should be postponed, at least until that time at which the prices have decreased considerably due to competition in the market, so as not to exacerbate the financial burdens for the less affluent members of society.

5.3 Conclusion

The analysis of currently available data suggests that more reliable information on the neurophysiological mechanisms of action of tDCS and TMS is necessary. Even though the physiological profile of these non-invasive neurostimulation devices seems to be safe in strictly controlled laboratory settings (i.e., with sufficiently trained users), if inadequately regulated, they can incur social and health risks.

With suitable modifications the Economic Disincentives Model, might provide a starting point for establishing long term physiological and social effects of tDCS. Furthermore, EDM for users of tDCS should be extended to TMS as a service. This unified "neurostimulation enhancement device use" license would enable citizens above the age of 25 to purchase and use tDCS devices and to benefit from the "service provider model" as applied to TMS. Only qualified medical doctors and neuroscientists could apply for a license to offer TMS as a service. Because of the complexity of the brain, it is questionable whether we will be able to overcome trade-offs between

[9]This could raise an issue of fairness for people who would like to benefit from enhancement services, but are unwilling or unable to demonstrate sufficient understanding of such complicated issues such as brain regions, electrical and magnetic fields, etc. Now, whether such an issue would materialize or not is an open question, but the provision that EDM would need to be fine-tuned (perhaps by dividing the exam in the theoretical and practical part, with one being sufficient to benefit from the service and both being necessary for product purchase) could offset this concern.

enhancement and concurrent impairment by enhancement devices (see Blank 2013), but citizens certainly have the right to pursue even potentially dangerous activities, as long as safety and quality control standards along with consumer protection and fair market practices are ensured.

References

Adee, S. 2012. Zap your brain into the zone: Fast track to pure focus. *New Scientist* 2850. http://www.newscientist.com/article/mg21328501.600-zap-your-brain-into-the-zone-fast-track-to-pure-focus.html?full=true. Accessed 23 Mar 2012.

Anonymous. 2013. Brain blast: DIY attempts at electrical brain stimulation to improve cognition are to get easier. *Nature* 498: 271–272.

Antal, A., et al. 2004. Facilitation of visuo-motor learning by transcranial direct current stimulation of the motor and extrastriate visual areas in humans. *European Journal of Neuroscience* 19: 2888–2892.

Arul-Anandam, A.P., C. Loo, and P. Mitchell. 2010. Induction of hypomanic episode with transcranial direct current stimulation. *Journal of Electro-Convulsive Therapy* 26 (1): 68–69.

Bersani, F.S., A. Minichino, P.G. Enticott, et al. 2013. Deep transcranial magnetic stimulation as a treatment for psychiatric disorders: A comprehensive review. *European Psychiatry* 28: 30–39.

Bikson, M., S. Bestman, and S. Edwards. 2013. Neuroscience: Transcranial devices are not playthings. *Nature* 501: 167.

Blank, R.H. 2013. *Intervention in the brain: Politics, policy and ethics*. Cambridge: MIT Press.

BMA. 2007. *Boosting your brainpower: Ethical aspects of cognitive enhancements. A discussion paper from the British Medical Association [BMA]*. http://www.bma.org.uk/ap.nsf/AttachmentsByTitle/PDFCognitiveEnhancement2007. Accessed 23 Mar 2012.

Boniface, S., and U. Ziemann (eds.). 2003. *Plasticity in the human nervous system: Investigations with transcranial magnetic stimulation*. Cambridge: Cambridge University Press.

Boroojerdi, B., M. Phipps, L. Kopylev, C.M. Wharton, L.G. Cohen, and J. Grafman. 2001. Enhancing analogic reasoning with rTMS over the left prefrontal cortex. *Neurology* 56: 526–528.

Bostrom, N., and A. Sandberg. 2009. Cognitive enhancement: Methods, ethics, regulatory challenges. *Science and Engineering Ethics* 15: 311–341.

Brunoni, A.R., et al. 2011a. Clinical research with transcranial direct current stimulation (tDCS): Challenges and future directions. *Brain Stimulation*. https://doi.org/10.1016/j.brs.2011.03.002.

Brunoni, A.R., et al. 2011b. Manic psychosis after sertraline and transcranial direct-current stimulation. *Journal of Neuropsychiatry and Clinical Neuroscience* 22 (3): e4–e5.

Clark, V.P., and R. Parasuraman. 2013. Neuroenhancement: Enhancing brain and mind in health and in disease. *Neuroimage* 85: 889–894.

Clark, V.P., et al. 2012. TDCS guided using fMRI significantly accelerates learning to identify concealed objects. *Neuroimage* 59: 117–128.

Cohen Kadosh, R., et al. 2010. Modulating neuronal activity produces specific and long-lasting changes in numerical competence. *Current Biology* 20: 2016–2020.

Cohen Kadosh, R., N. Levy, J. O'Shea, et al. 2012. The neuroethics of non-invasive brain stimulation. *Current Biology* 22 (4): 108–111.

Dockery, C.A., R. Hueckel-Weng, N. Birbaumer, and C. Plewnia. 2009. Enhancement of planning ability by transcranial direct current stimulation. *Journal of Neuroscience* 29 (22): 7271–7277.

Dockery, C.A., D. Liebetanz, N. Birbaumer, M. Malinowska, and M.J. Wesierska. 2011. Cumulative benefits of frontal transcranial direct current stimulation on visuospatial working memory training and skill learning in rats. *Neurobiology of Learning and Memory* 96 (3): 452–460.

Dubljević, V. 2012a. Principles of justice as the basis for public policy on psycho-pharmacological cognitive enhancement. *Law, Innovation and Technology* 4 (1): 67–83.

Dubljević, V. 2012b. Toward a legitimate public policy on cognition-enhancement drugs. *American Journal of Bioethics – Neuroscience* 3 (3): 29–33.
Dubljević, V. 2013. Autonomy in neuroethics: Political and not metaphysical. *American Journal of Bioethics – Neuroscience* 4 (4): 44–51.
Dubljević, V. 2015. Neurostimulation devices for cognitive enhancement: Toward a comprehensive regulatory framework. *Neuroethics* 8 (2): 115–126. https://doi.org/10.1007/s12152-014-9225-0.
Dubljević, V., V. Saigle, and E. Racine. 2014. The rising tide of transcranial Direct Current Stimulation (tDCS) in the media and academic literature. *Neuron* 82: 731–736.
EGE. 2005. *Opinion on the ethical aspects of ICT implants in the human body* (16 March 2005). *Opinion of the European Group on ethics in sciences and new technologies to the European Commission*, vol. 4. Luxembourg: Office for Official Publ. of the European Communities.
Fitz, N.S., and P.B. Reiner. 2013. The challenge of crafting policy for do-it-yourself brain stimulation. *Journal of Medical Ethics*, Epub ahead of print 3 June 2013. https://doi.org/10.1136/medethics-2013-101458.
Fitzgerald, P.B., and Z.J. Daskalakis. 2013. *Repetitive transcranial magnetic stimulation treatment for depressive disorders: A practical guide*. Dodrecht: Springer.
Fregni, F. 2005. Transcranial direct current stimulation. *British Journal of Psychiatry* 186: 446–447.
Galvez, V., et al. 2011. Hypomania induction in a patient with bipolar II disorder by transcranial Direct Current Stimulation (tDCS). *Journal of Electro-Convulsive Therapy* 27: 256–258.
Hamilton, R., et al. 2011. Rethinking the thinking cap: Ethics of neural enhancement using noninvasive brain stimulation. *Neurology* 76 (2): 187–193.
Illes, J., and B. Sahakian (eds.). 2011. *Oxford handbook of neuroethics*. Oxford: Oxford University Press.
Iuculcano, T., and R. Cohen Kadosh. 2013. The mental cost of cognitive enhancement. *The Journal of Neuroscience* 33 (10): 4482–4486.
Jacobson, L., M. Koslowski, and M. Lavidor. 2012. tDCS polarity effects in motor and cognitive domains: A meta-analytical review. *Experimental Brain Research* 216: 1–10.
Keck, M.E., I. Sillaber, K. Ebner, et al. 2000. Acute transcranial magnetic stimulation of frontal brain regions selectively modulates the release of vasopressin, biogenic amines and amino acids in the rat brain. *European Journal of Neuroscience* 12: 3713–3720.
Kincses, T.Z., et al. 2003. Facilitation of probabilistic classification learning by transcranial direct current stimulation of the prefrontal cortex in the human. *Neuropsychologia* 42: 113–117.
Larriviere, D., M.A. Williams, M. Rizzo, and R.J. Bonnie. 2009. Responding to requests from adult patients for neuroenhancements: Guidance of the ethics, law and humanities committee. *Neurology* 73 (17): 1406–1412.
Luber, B., and S.H. Lisanby. 2013. Enhancement of human cognitive performance using transcranial magnetic stimulation (TMS). *Neuroimage*, [Epub ahead of print] 13 June 2013. https://doi.org/10.1016/j.neuroimage.2013.06.007.
Luber, B., J. Steffener, A. Tucker, C. Habeck, A.V. Peterchev, Z.D. Deng, R.C. Basner, Y. Stern, and S.H. Lisanby. 2013. Extended remediation of sleep deprived-induced working memory deficits using fMRI-guided transcranial magnetic stimulation. *Sleep* 36 (6): 857–871.
Maguire, G.Q., and E.M. McGee. 1999. Implantable brain chips? Time for debate. *Hastings Center Report* 29 (1): 7–13.
Maslen, H., J. Savulescu, T. Douglas, N. Levy, and R. Cohen Kadosh. 2013a. Regulation of devices for cognitive enhancement. *The Lancet* 382: 938–939.
Maslen, H., T. Douglas, R. Cohen Kadosh, N. Levy, and J. Savulescu. 2013b. Do-it-yourself brain stimulation: A regulatory model. *Journal of Medical Ethics*, E-pub ahead of print, https://doi.org/10.1136/medethics-2013-101458.
Maslen, H., T. Douglas, R. Cohen Kadosh, N. Levy, and J. Savulescu. 2014. The regulation of cognitive enhancement devices: Extending the medical model. *Journal of Law and the Biosciences* 1 (1): 68–93.
McKinley, R.A., N. Bridges, C.M. Walters, and J. Nelson. 2012. Modulating the brain at work using noninvasive transcranial stimulation. *NeuroImage* 59: 129–137.

Pascual-Leone, A., F. Fregni, M. Steven-Wheeler, and L. Forrow. 2011. Non-invasive brain stimulation as a therapeutic and investigative tool: An ethical appraisal. In *Oxford handbook of neuroethics*, ed. J. Illes and B. Sahakian, 417–440. Oxford: Oxford University Press.

Patel, P., C.C. Meltzer, H.S. Mayberg, and K. Levine. 2007. The role of imaging in United States courtrooms. *Neuroimaging Clinics of North America* 17 (4): 557–567.

Poreisz, C., et al. 2007. Safety aspects of transcranial direct current stimulation concerning healthy subjects and patients. *Brain Research Bulletin* 72: 208–214.

Priori, A. 2003. Brain polarization in humans: A reappraisal of an old tool for prolonged non-invasive modulation of brain excitability. *Clinical Neurophysiology* 114: 589–595.

Ranisch, R., D. Garofoli, and V. Dubljević. 2013. 'Clock shock,' motivational enhancement and performance maintenance in Adderall use. *American Journal of Bioethics – Neuroscience* 4 (1): 13–14.

Richter, L. 2013. *Robotized transcranial magnetic stimulation*. Dodrecht: Springer.

Riedel, P., et al. 2011. Contact dermatitis after transcranial direct current stimulation. *Brain Stimulation* 5 (3): 432–434.

Rioult-Pedotti, M.S., and J.P. Donoghue. 2003. In *Plasticity in the human nervous system: Investigations with transcranial magnetic stimulation*, ed. S. Boniface and U. Ziemann, 1–25. Cambridge: Cambridge University Press.

Rossi, S., M. Hallett, P.M. Rossini, et al. 2009. Safety, ethical considerations, and application guidelines for the use of transcranial magnetic stimulation in clinical practice and research. *Clinical Neurophysiology* 120 (12): 2008–2039.

Rothwell, J.C. 2003. Techniques of transcranial magnetic simulation. In *Plasticity in the human nervous system: Investigations with transcranial magnetic stimulation*, ed. S. Boniface and U. Ziemann, 26–61. Cambridge: Cambridge University Press.

Sample, I. 2010. Electrical stimulation of the brain boosts math skills, claim scientists. *The Guardian*, 4 Nov 2010.

Sandberg, A., W. Sinnott-Armstrong, and J. Savulescu. 2011. Cognitive enhancements in court. In *The Oxford handbook of neuroethics*, ed. J. Illes and B. Sahakian, 273–284. Oxford: Oxford University Press.

Schermer, M. 2014. Reducing, restoring or enhancing autonomy with neuromodulation techniques. In *Free will and the brain: Neuroscientific, philosophical, and legal perspectives*, ed. W. Glannon. Cambridge: Cambridge University Press.

Science and Technology Options Assessment [STOA]. 2009. *Human enhancement study*. The Hague: Rathenau Institute.

Simpson, K.N., M.J. Welch, A.F. Kozel, M.A. Demitrack, and Z. Nahas. 2009. Cost-effectiveness of transcranial magnetic stimulation in the treatment of major depression: A health economics analysis. *Advances in Therapy* 26 (3): 346–368.

Snyder, A. 2009. Explaining and inducing savant skills: Privileged access to lower level, less-processed information. *Philosophical Transactions of the Royal Society of Britain* 364: 1399–1405.

Snyder, A., H. Bahramali, T. Hawker, and D.J. Mitchell. 2006. Savant-like numerosity skills in normal people by magnetic pulses. *Perception* 35: 837–845.

Stagg, C.J., and M.A. Nitsche. 2011. Physiological basis of transcranial direct current stimulation. *Neuroscientist* 17 (1): 37–53.

Sutherland, S. 2013. Magnetic brain stimulation could ease pain. *Scientific American*, 17 Mar 2013. http://www.scientificamerican.com/article.cfm?id=magnetic-brain-stimulation-could-ease-pain. Accessed 15 Dec 2013.

United Nations [UN]. 1971. *Convention on psychotropic substances*. www.unodc.org/pdf/convention_1971_en.pdf. Accessed 3 Apr 2013.

Warwick, K. 2008. Cybernetic enhancements. In *Reshaping the human condition: Exploring human enhancement*, ed. L. Zonneveld, H. Dijstelbloem, and D. Ringoir, 123–131. The Hague: Rathenau Institute.

Part III
Objections and Responses: Public Policy and Autonomy Revisited

Chapter 6
Is the Proposal of the Political Notion of Autonomy Problematic?

My prior work on autonomy (see Dubljević 2013 and Chap. 3) took up the challenge posed by Felsen and Reiner (2011) to substantially address how autonomy should be viewed in light of new evidence from (cognitive) neuroscience. I argued that it is premature to propose that the empirical data renders autonomy 'Quixotic', posited that the moral–political notion of autonomy was mistakenly associated with the metaphysical concept of "free will", and offered ideal-typical degrees of coercion and compulsion which qualify instances in which autonomy might be diminished or entirely lacking. However, my argument caused some controversy in terms of many objections, and some support for the position that autonomy should be viewed as a gradual capacity (Nagel 2013) and a right (Dolan 2013). Given the many-faceted and sometimes completely opposing positions that inform these objections, I will refute them one by one.

6.1 Is the Capacity for Autonomy Insufficient?

Felsen and Reiner (2013) point out several areas in which they have reason to disagree with my views on autonomy. The most important point of contention is that, unlike my proposal, which emphasizes rights and underlying capacities, as opposed to actual engagement of capacities and enjoyment of rights, Felsen and Reiner "maintain that what our brains *actually do* when they make decisions is more practically relevant than what they are capable of doing" (p. 52). Felsen and Reiner also find faults with my qualitative designation of degrees of coercion, and offer a quantitative rendition of the criterion for undue coercion. They suggest that "influences that make a particular decision only slightly more likely (5–10%, for example) are not undue"

This chapter draws and substantially expands on my previous work which has been published as: Dubljević (2016).

V. Dubljević, *Neuroethics, Justice and Autonomy: Public Reason in the Cognitive Enhancement Debate*, The International Library of Ethics, Law and Technology 19,
https://doi.org/10.1007/978-3-030-13643-7_6

whereas "influences that are highly likely (90–95%, for example) to lead to a particular decision should be considered undue, *even though the individual retains the capacity to choose otherwise*" (p. 53, emphasis in the original). Finally, they insist the process of authentication is relevant for the incorporation of cognitive content which has originated outside of the cranium of the agent in question.

So let's start with the issue of authentication. In my work, I use the term endorsement, not only because it meshes well with the Rawlsian character of my proposal, but also because it allows for a conceptual split with the traditional hierarchical concept of autonomy (e.g., Frankfurt 1971, 1988), which has some metaphysical baggage attached to it. The term endorsement avoids the vagueness of the term authenticity, and at the same time stresses the social character and justification of processes in question. The key is to understand that endorsability is not entirely subjective, but constrained with mutually acceptable terms of social cooperation. Valuable philosophical work, not just by Rawls (2005), but also by Kitcher (2011) specifies this social character of endorsability. For instance, Kitcher argues that certain goals are endorsable "just in the case there are possible environments in which they could be satisfied for all our fellows" (Kitcher 2011, p. 223). Kitcher's examples of wanting to have adequate food as socially endorsable versus wanting to prevent all others from reproducing as socially unendorsable are, I think, very important for complementing what Felsen and Reiner call 'principled suasion' as due form of external influence. Namely, we can imagine a group which indoctrinates certain individuals to firmly believe they are entitled to prevent others from reproducing. Thus, the endorsability criterion serves us better in filtering out influences which are detrimental for autonomy and which are not.

This leads to the second point. Even though Felsen and Reiner are right to point out that qualitative ideal-typical forms of coercion might have more traction if they were translated as quantitative designators of stochastic influence, it is not just the degree (or percentage) that matters. Indeed, direct coercion as I define it (external influences backed by sanctions), if they are socially justified in terms of legal responses to undue influences for personal gain of equal degree (e.g., mugging), should not be considered undue even if their power is such that it is highly likely (90–95%) to actually influence certain behavior. The point is that laws against robbery are socially endorsable and do not diminish autonomy in any way, whereas robbery itself diminishes autonomy of others, even if an agent has expressed a genuine desire to get robbed in order to, say, prove their courage.

This leaves the issue of capacity versus actuality, as crucial for autonomy. I have argued that the mere fact that a person has a right to a course of action (e.g., to ask for a second opinion in medical settings), and relevant volitional and cognitive capacities in absence of undue coercion or compulsion to follow through with such a course of action, is sufficient for the basic idea of autonomy. Now, Felsen and Reiner could be understood to assume that since decision making is stochastic, if a person has declined to ask for a second opinion in a sufficient number of medical encounters, this might signal the fact that the likelihood that they would ever ask for a second opinion is negligible and thus their autonomy might be put into question. This is a reasonable and serious worry. However, even though we might agree that such a person would be far from the ideal of autonomy, basic autonomy which allows for the right to ask for a second opinion could still be presumed in absence of clear pathology. The endorsability criterion sheds some light on why this might be the case once again. Namely, we could imagine Mr. X, who loves his wife very much and deeply cares about her opinion. He regularly asks her advice and commits to the course of action she suggests without consideration 95% of the time, 4% of the time, he actually takes the time to consider the advice, and 1% of the time he rejects it without consideration. Now, Mr. X's actions are entirely endorsable, since in Kitcher's terms there are possible environments in which the goals could be satisfied for all others as well. This grounds the basic autonomy of Mr. X's actions and guards against stripping Mr. X of his rights to make different decisions should circumstances change (e.g., he finds out his wife has been cheating on him, or spending their joint assets to the benefit of her high school sweetheart).

For those that are (perhaps like Felsen and Reiner) less swayed by thought experiments and more by empirical backing, another, real life example might be useful. For instance, in Nigeria patients often relinquish not only their right for a second opinion, but also their consent power to the treating physician. It would be easy to conclude that the majority do not wish to have the power of consent in the first place. However, this conclusion would be very wrong, not just in normative terms, but also in terms of what these patients actually wish. Adeleye and Tobin (2013) showed how the right to consent is very important to people who will ultimately choose to waive it: out of 200 hospitalized patients interviewed in a Nigerian community, 92.5% indicated that their consent is necessary.

To sum up, the translation of degrees of coercion into quantitative terms (e.g., indirect coercion makes certain decisions 10% more likely) might be useful. However, the endorsability criterion is much better at designating autonomous action than 'authentication', and there are good reasons to maintain the view that capacity and rights are more important than actual decisions made or the exact stochastic likelihood of influencing a certain decision.

6.2 Is the Political Account of Autonomy an Unjustified Theory?

Some objections, like those of Levy (2013), actually question the theoretical underpinnings of the political view of autonomy. Levy asserts that "[j]ust as the pragmatists hoped to avoid theory, so Veljko Dubljević hopes to avoid metaphysics" (p. 53). According to Levy what "pragmatism gives us is just another theory: not a way of escaping from theory and the need for theoretical justification, but a theory that needs justification as much as any other" (p. 53). So, has the hard theoretical work for the political conception of autonomy been done? Although more can certainly be said, the theoretical foundations are already fairly established in pragmatist (Dewey), analytical (Rawls), and continental philosophy (Habermas).

Dewey (1921), for example, offered a theoretical account of a non-metaphysical view of freedom. Dewey defined three relevant conditions: "(i) efficiency in action, ability to carry out plans, the absence of cramping and thwarting obstacles, (ii) ... capacity to vary plans, to change the course of action, to experience novelties ... and (iii) ... the power of desire and choice to be factors in events" (303–4). Now, even though Dewey was not as clear regarding terminology (autonomy versus freedom) and empirical observability of relevant components, he did distinguish between freedom as capacity and as right supplanted by social agreement (307), view coercion as pervasive and in need of justification (308) and insisted on the importance of planning for future and control (311–313).

Rawls' theory is situated in analytical philosophy and provides a robust theoretical foundation for viewing freedom in political and not metaphysical terms (see Rawls 2005, 29ff. and n. 31). According to Rawls, citizens are "free" in three aspects:

1. "They conceive of themselves and of one another as having the moral power to have a conception of the good" (30).
2. "They regard themselves as being entitled to make claims on their institutions so as to advance their conceptions of the good (provided these conceptions fall within the range permitted by the public conception of justice)" (32).
3. "They are viewed as capable of taking responsibility for their ends and this affects how their various claims are assessed" (33).

In Rawls's theory, autonomy and rights are socially constituted, and not inherent properties established by any "essence." The difference between a free individual and a slave is one of rights, not of capacities.

However, Rawls's theory is not an isolated case. Other theories delve more deeply into historical changes from traditional to democratic societies in order to establish the normative justification for autonomy. In Habermas's political thought, which is situated in continental philosophy and the Frankfurt school of critical theory, the key issue is the project of realizing the system of rights that would guarantee private and public autonomy "under the conditions of *postmetaphysical thinking*" (Habermas 2004, 443, emphasis mine). According to Habermas, the democratic outlook established with the French Revolution was "expressed in the conviction that emancipated individuals are jointly called to be authors of their destiny. In their hands lies the power to decide about the rules and manner of their living together. As citizens, they give themselves the laws they want to obey, thereby producing their own life context" (p. 468). Furthermore, it "was expressed [...] in the conviction that the exercise of political domination could be legitimated neither religiously (by appeal to divine authority) nor metaphysically (by appeal to an ontologically grounded natural law). From now on, a politics radically situated in this world should be justifiable [...] using the tools of *postmetaphysical theorizing*" (469, emphasis mine).

Even though, for a lack of space, I cannot delve deeper into the theoretical foundations, my assertion that autonomy and rights are based on democratic authority and empirically observable capacities, and not metaphysical notions such as "free will," has a fairly established theoretical underpinning. There might be a need to further refine the definition of "degrees of autonomy," make the terminology more clear (e.g., decision making, not choosing), or specify instances in which autonomy might be diminished, but there is no need to further defend the theoretical feasibility of the approach for neuroethics.

6.3 Is the Political Approach to Autonomy Too Demanding or Nothing New?

Rather than assert that there is not enough theoretical background for the approach, some objections assert that old compatibilist theories of 'free will' actually provide all the content for the political approach of autonomy. Namely, Sharp and Wasserman (2013) state that my analysis "has similarities to several compatibilist approaches" (p. 55), such as Frankfurt's. They also find faults with the endorsability criterion, first stating that explicit endorsement of actions is too demanding, and that what they see is the weaker 'potential endorsability' criterion which they think "alters an untenable position" (p. 55), and therefore runs into problems of its own.

So, let's start with clarifying the point about endorsing. It should be emphasized that endorsing comes in many forms: one can endorse an action explicitly, for instance saying 'I did this;' one can explicitly disown an action, for instance by apologizing after, say, spilling wine by accident; and one can implicitly endorse an action through further actions. So let's see if this clarification helps put some of Sharp and Wasserman's concerns to rest. They first offer an example of the "wanton" who may do terrible harm to others based on whim, without endorsing the harm. According to Sharp and Wasserman, my account would exculpate such a person. But this is clearly not the case, since the implicit endorsability criterion shows that the wanton's actions are led by the principle 'I do as I please' at different times. Thus, even though he denies endorsing the harm, his actions speak for themselves, and he is deemed responsible for endorsing a reckless life-plan that is bound to cause harm.

How does their second example fare? Sharp and Wasserman think that the criterion of potentially continuing to endorse an action after a period of critical informed reflection is too demanding and offer the case of making certain impulsive decisions which we regret afterwards. They state:

> there are many decisions I [sic] might make of my own volition—say, to rob banks—that I could not continue to make after informed critical reflection. I might come to see that the risks of bank robbery clearly outweigh the benefits [...]. [...] Yet despite the fact that I could not endorse my behavior after sustained reflection, I should not be exculpated for the harms I cause if I decide, on an unreflective whim, to follow the path of Bonnie and Clyde. I ought to be held accountable. (p. 56)

Now, this objection can be answered by bearing in mind the distinction between *explicit* and *implicit* endorsement and *disowning* of actions, which I introduced above. Sharp and Wasserman are of course right to point out that we make many actions that we regret afterwards, once we have the time for critical reflection. But autonomy and responsibility are imputed by default: we take great pains to *explicitly disown* some actions that we regret, for instance by saying that we didn't mean to, and by promising not to do that sort of transgression again. We might be believed and excused the first time (if the transgression is minor), but repeating the action would mean endorsing it as part of our life-plan. At other times, we recognize that certain actions are very wrong, and try to hide the fact that we had done it, and at the same time, we *implicitly* endorse that we have done something we know is very wrong. The bank robber who has a change of heart and tries to abandon the life of crime while covering his tracks would be considered fully autonomous and responsible on the view of autonomy that I defend.

Finally, the charge that the political account of autonomy is simply re-packaged compatibilism should be briefly addressed. It is true that certain elements are common to both certain metaphysical positions and the political conception of autonomy. But this is merely the result of 'overlapping consensus' (see Rawls 2005). Namely, the political conception of autonomy assumes some common elements from the background political culture, which is informed by comprehensive metaphysical doctrines. When I say that the political conception of autonomy neither presupposes nor denies compatibilism, (or libertarianism or hard determinism, for that matter) I am implying that like Rawls' conception of justice, and autonomy could be seen as reasonable by proponents of any of those views. For instance, even though hard determinists deny that there is anything like free will (see e.g., Greene and Cohen 2004), they do not deny that there should be some reasonable criterion for when some people should be deprived of their socially recognized liberty and incarcerated. Similar denial of certain doctrinary elements and acceptance of reasonable social practices holds for other metaphysical positions as well. Thus, the crux of the argument is that there are elements of social practices that are (or could be) recognized by proponents of any of those comprehensive doctrines as reasonable, and this has been iterated by numerous philosophers in the past. I take Sharp and Wasserman's comments as proof that proponents of compatibilist doctrine might see the political conception of autonomy as reasonable, at least when properly explained.

6.4 Is Post Hoc Endorsing an Illusion?

Hui (2013) objects to my position by asserting that a "theory of autonomy that relies heavily on post hoc rationalizations is problematic, since choices often exist in some gray territory of being incompletely free and can usually be rationalized one way or another" (p. 57). She thinks that unconscious influences such as framing of information in clinical encounters might present a profound challenge for autonomy, and by invoking the example of split-brain patients and the 'spreading of alternatives effect', contends that merely allowing endorsement after the fact would make many intuitively unfree actions autonomous, which could be problematic. Finally, Hui echoes Felsen and Reiner by emphasizing actual use of cognitive capacities, but goes a step further and states that agents "must actively exercise this capacity *before* they act in order for it to give rise to any meaningful notion of autonomy" (p. 58).

First of all, I agree with Hui on the potential detrimental effects of framing in clinical encounters. As I have argued at length elsewhere, trust in physicians is only instrumental for obtaining protection of the patient's best interests (Dubljević 2015) and any abuse of this trust by framing or nudging practices needs to be curbed (Bell et al. 2013). It is very important that full and adequate information is provided during the informed consent process (Bryan and Dubljević 2017), so that competent adults can make an informed choice, *should they so choose* (see the discussion on flipping a coin below).

That said, Hui's proviso that each action has to be thought out in advance in order to count as autonomous is clearly too demanding. The arguments put forward by Sharp and Wasserman, to which I responded above, could be echoed here to blunt the sting of Hui's challenge. But what of the split-brain patient challenge? A careful reading of my view of autonomy would show that we can avoid at least some counter-intuitive results that Hui fears. Namely, split-brain patients have a clear pathology, and that is why we might consider their actions to be less autonomous, which is entirely consistent with my view. Now, what about the 'spreading of alternatives effect'—the tendency to evaluate the chosen alternative more positively and the rejected alternative more negatively than before the decision? Hui worries that certain actions would be endorsed after they are chosen, even though equally good actions could have been chosen as well.

To this I respond by pointing to the prerequisites of reasonableness, more specifically to the "willingness to recognize burdens of judgment" (Rawls 2005, p. 54). These are in Rawls' words some of the "many hazards involved in the correct (and conscientious) exercise of our powers of reason and judgment in the ordinary course of political life" (p. 56). Psychology, cognitive science and neuroscience are constantly updating the range of hazards that affect our everyday judgment. And yet, most of the time, we are well equipped to deal with them, and erring on the side of (presuming) autonomy is not something that would cause great concerns in the political sense. Consider an example of Mr. X once again. He was the husband who would normally follow the advice of his wife. As I argued above, his decision to follow the advice of a person that he believes is in a good epistemic condition, and has his best interest in mind, is perfectly autonomous. Now, let's say that he is divorced from his wife and, being the lazy fellow that he is, he opts to make some of his binary decisions by tossing a coin. Saying that the other option might have been a good one (or even better) does not diminish the autonomy of his choice to simply follow an easy decision strategy. He might regret this decision, and might disown making important decisions in this manner in the future. Regardless, implicit endorsement of following through with his decision, backed up undoubtedly by psychological mechanisms of goal preservation, which sometimes might skew his perception, is perfectly normal and does not reduce autonomy in any way.

6.5 Is There a Need to Entertain Metaphysics in the Discussion About Addiction?

Some objections focus on the practical issues of policy, and the scientific theory that should be informing it. For instance, van der Eijk (2013) argues that "a dialogue between addiction metaphysics and policy should continue, because this dialogue is not the source of the problems regarding 'addict autonomy'" (p. 58). She quotes psychosocial theories of addiction to argue that "addictive drug use may be considered as a means of coping with a constricting social environment: the only pleasurable option left in an otherwise stressful or meaningless life" (p. 59). Van der Eijk opposes what she sees as 'in between' mix of 'moral choice' and 'brain disease' models of addiction and argues that addiction really is a psychological coping mechanism reacting to environmental influences.

I begin by emphasizing that van der Eijk and I are in agreement in many respects—we both call for more effective socio-political interventions for treatment of harmful addictions, and maintain that some of our benign habits can be seen as addictions (I gave the example of oxygen, van der Eijk opted for a more socially embedded instance of romantic relationships), and that even though they affect our decision making processes profoundly, these habits nevertheless do not reduce autonomy. If the example of Mr. X above is remembered, his loving relationship with his wife influenced his decision making in profound ways, and yet he is still considered fully autonomous, and retains the right to change his mind (should the information of the wife's disloyalty become known to him, or for any other reason he might find important). That said, I have to emphasize that addiction is only a test case for the political conception of autonomy—there is no presumption that well established (and conflicting) theories of initiation, continuation and relapse of self-harming use (and behavior) will be displaced, or somehow adjudicated by this work. In this sense, van der Eijk's advocacy for a specific theory is a legitimate position, but does not force my view to conform to any single theory within the debate among National Institute on Drug Abuse (NIDA) recognized theories. Namely, the psychosocial theory of addiction, which van der Eijk favors, is only one competing theory in the cluster of theories that focus on the individual's relationship with society. Within that group there are quite a few theoretical viewpoints on drug abuse and addiction, and of course, there are additional clusters of theories such as theories based on the relationship with self; significant others; or nature (see Lettieri et al. 1980). The political conception of autonomy leaves the issue of specific psychological, physiological and environmental effects to the scientific debate and starts from only very general presuppositions that are supported by the majority of theories.

In this sense, I find it regrettable, if understandable, that van der Eijk chooses to call for more and not less metaphysics. Now, van der Eijk is correct to assert that metaphysical and even religious considerations (in terms of social support structure) figure in individual decision making pertaining to a range of phenomena, including addiction. However, I have to strongly re-emphasize that policies and laws should be grounded on rights, and not metaphysical notions individuals might subscribe to.

That said, I acknowledge that rights are only part of political identity of citizens in democratic societies, and that citizens certainly have metaphysical and religious convictions. Therefore, the fact that neuroscience and debate about addiction have been coupled with metaphysics is not problematic per se, but, as I have argued, is misplaced at the level of practically ascribing moral and legal responsibility in liberal democracies. It should be noted that the basis for ascription of moral and legal responsibility differs in traditional and pluralist democratic societies. In a traditional society, the normative force of the legal system is based on a metaphysical "truth" or religious authority. However, liberal democracies are supposed to be legitimate for citizens upholding diverse world-views and theoretical positions, and as such neither presuppose nor are hostile toward any reasonable metaphysical (or religious) doctrines. Thus, in contemporary democratic states, laws and policies start from observable empirical facts about relevant capacities (see Morse 2007). Of course, less democratic or tolerant states might have a proclaimed official metaphysical position (e.g., determinism/dialectical materialism in North Korea) or official religion (e.g., Shia Islam in Iran) and base their legal system accordingly. The fact of pluralism in western democracies prevents any single doctrine from serving as the basis for law and policy, and so reasonable conceptions of justice (see Rawls 2005) and autonomy (see Chap. 3 above) are the norm, rather than an exception.

6.6 Is There a Need for a Strong Universal Theory on How Autonomy Is Developed?

As mentioned earlier, some objections in fact second several of my conclusions, and raise some additional issues that they think should be considered. Nagel (2013), while supporting my distinction between metaphysical 'free will' and political autonomy, argues that binary distinctions arising in the law, such as the "all-or-none view of autonomy" (p. 61) need to be supplanted with autonomy assessment and support throughout childhood, adolescence and adult life, especially in health care settings. Nagel argues that development is a "crucial aspect of autonomy that deserves further clarification" (p. 60) both in terms of gradual maturation during childhood and adolescence (as recognized by 'anticipatory autonomy rights' or 'right to an open future'), and the gradual fostering of the capacity for mature self-governance in adults facing loss of cognitive ability and depletion of self-control in non-pathological states.

Since Nagel and I are very much in agreement on many issues, I will only briefly remark on the need for binary distinctions and the role of gradual capacities as opposed to counter-intuitive metaphysical problems. As I have argued in Chap. 3 and above, in liberal democracies, the law is a normative structure that draws on moral–political concepts and makes binary distinctions that are grounded in specific gradual capacities, but sharp by necessity. In a sense, these are arbitrary and 'socially constructed,' but apart from the necessity to define the degrees of a relevant capacity, this in no way makes it obsolete (or 'metaphysical' for that matter). Consider the case of maturity: it is vague, gradual, and linked with a variable ability—self-control. Conceptual and empirical analysis might provide relevant insights, but the exact age at which persons are entitled to, say, drink, drive, vote, and freely dispose of their property is a social construct that differs from one society to another. In some societies the age is 18 years for all these relevant rights (e.g. Germany), while no amount of analysis would provide definite answers to what is so drastically different between persons aged say, 17 years and 11 months, and 18 years and 1 day. In other countries (e.g., most of the United States), the ages are 21, 16, and 18 years, respectively, although one might find individuals that are mature enough to vote at 16 (and Austria grants voting at 16), and those that are quite immature at 21 (the voting age for men in Saudi Arabia).[1] That is not to say that this is entirely arbitrary. Societies create sharp boundaries where in reality there are merely gradual changes, but there are legitimate reasons for consistent boundaries to be set.

Now, if autonomy was not a right based on social agreement, but a mysterious capacity stemming from "free will," something odd would be happening: at the age of 18, "free will" would spontaneously materialize, where there had been none. Critics of my view would be right to point that historically law used to rely on "free will" and other metaphysical and/or religious doctrines. However, with the advent of democracy, autonomy and rights are based on democratic authority and empirically observable capacities, not metaphysical or religious doctrines. This is not to say that discussions about neuroscience, addiction and free will are irrelevant *tout court*—only that they are not necessary for establishing moral and legal responsibility, and that they are a poor guide for policy.

Now, Nagel and I agree on this, and the need to "determine crucial thresholds that demarcate moral responsibility" (Nagel 2013, p. 61). However, I think I have made abundantly clear that assessment of autonomy can only be useful in particular environments (Nagel is correct to point health care setting as one of them) in which social constructions dictated by fair terms of cooperation are set. Specific assessments will undoubtedly come up with only qualified (and qualitative) measures of autonomy, and various societies will emphasize different levels of development and functioning as the basis for certain rights. Thus, I fear no single universal threshold across societies will be forthcoming: there might only be a reasonable range of thresholds that is accepted by pluralistic democratic societies as a result of public debate and deliberation.

[1] See "Legal voting age around the world," available at: chartsbin.com/view/re6, accessed December 20, 2014.

6.7 Is the Political Conception of Autonomy Relying on Social Psychology Evidence at the Detriment of Neuroscience?

Horstkotter and Snoek (2013) offer a poignant critique of what they take to be the "nontransparent way in which [Dubljević] either rejects or embraces certain scientific findings" (p. 62) and the 'heteronomy' of rational life-plans that conform to social and external factors. Horstkotter and Snoek argue that my view tries to avoid complex philosophical discussions engendered by neuroscientific evidence by rejecting the relevance of neuroscientific studies altogether and leaning on "antagonistic strands of social psychology studies that invoke concepts of self-control and emphasize the reality and significance of autonomous human behavior" (p. 62). Apart from making the accusation of "invoking whatever is consistent" with my view and "rejecting everything inconvenient" (p. 62), Horstkotter and Snoek take aim at the idea of 'rational life-plans' which are shaped by fair terms of social cooperation. They argue that "people's decisions to act should not only be deliberated on, but as a result also be an expression of what they value and who they consider themselves to be" (p. 63).

First of all, these objections make me keenly aware that some of my formulations were open to misunderstanding. Perhaps it needs to be made more clear that I am in no way rejecting neuroscientific evidence: as long as the methodology is sound, every set of replicable data, whether it comes from neuroscience or social psychology, gives us a welcome insight into the inner 'nuts and bolts' of human behavior and decision making. That said, data which is gathered and then reported can be interpreted in equally plausible ways, and the overall significance for society is subject to even more interpretation, which may or may not be backed by scientific evidence. Since Horstkotter and Snoek mention Wegner's "Illusion of conscious will" (2002), I will stick to this example to make my view clear: the scientific experiments cited in Wegner's work have undoubtedly shown that many actions and behaviors which were previously taken to be conscious and consciously initiated are in fact unconscious. Also, there is no doubt that there are psychological mechanisms at play which can be tricked to yield specific responses in controlled laboratory settings. These are the data, with which I have no qualms. However, Wegner's (and others') *interpretation of data* is that virtually no action is conscious or guided by reason, which leads to a complex philosophical discussion regarding the downstream social repercussions. Now, this is something very familiar in the history of philosophy: for instance, the early work in optical illusions has without a shadow of a doubt shown that there are mechanisms at play in the human visual system, which can be consistently tricked to yield specific responses. This has produced different interpretations of the data, and complex philosophical discussions regarding the downstream repercussions, which led Descartes famously to doubt that anything perceived by the senses is in fact real. Even though Descartes' complex philosophical discussion is very valuable in conceptual terms, it is by no means useful for practical purposes in real life settings. The human visual system might sometimes be fooled by different optical illusions, but hardly anyone denies that it serves us well in the mundane environments for which it has evolved.

So, what do I reject in Wegner's work? Only his interpretation that conscious deliberation is a faulty illusion. On the other hand, this evidence clearly supports fallibilism: errors in conscious judgment certainly exist but this does not mean that it is faulty, merely that it is fallible. Indeed, the capacity to consciously reason is indispensable for autonomy, and Wegner's data merely qualifies what can be expected (I will come back to this point below), and does not and cannot prove actions based on conscious reasoning to be illusory. This point is fairly established in the scientific community and as such succinctly formulated by Steven Pinker:

> The indispensability of reason does not imply that individual people are always rational or are unswayed by passion and illusion. It only means that people are *capable* of reason, and that a community of people who choose to perfect this faculty and to exercise it openly and fairly can collectively reason their way to sounder conclusions in the long run. (Pinker 2011, p. 181, emphasis in the original)

So, instead of rejecting data that is inconvenient, my approach is to consistently take the *data* very seriously and to subject the *interpretation of data* to critical scrutiny. When I said in Chap. 3 that the spill-over effect of metaphysics to practical philosophy was mistaken, I stand by my assertion: there is nothing wrong in having complex philosophical discussions regarding downstream implications of a specific interpretation of data, but there is a problem in conflating a singular (and implausible at that) interpretation of data as the data itself, and advocating for poorly thought out social policies based on cheap tricks—in other words, using the 'bait and switch' tactic: promising neuroscientific evidence based on high technology and delivering low grade metaphysics.

Coming back to Horstkotters and Snoek's worry about the rational life-plans: the actions people make autonomously need not be deliberated upon as long as they are guided by rational life-plans, or what they call an 'expression of what they value.' But actions are sometimes committed that do not conform to people's considerations of who they are. In some instances, people employ self-delusions to continue with their self-image in spite of contrary actions. The crux of the 'social' grounding of rational life-plans, which Horstkotter and Snoek think leads to heteronomy, is that only sustained critical scrutiny from others allows us to penetrate some of the self-serving delusions and leads to genuine autonomy. In this sense, Horstkotter and Snoek expect too much and too little from autonomy: they think individuals will always deliberate before making an action, and they posit that whimsical decisions, without consideration of others or the fair norms of social cooperation are autonomous.

6.8 Is the Political Conception of Autonomy Helpful in Practical Situations?

Similarly to Horstkotter and Snoek, some critics question the definition of endorsable rational life-plans and worry that conceiving of "rationality in terms of nondestructive long-term life-plan unduly restricts the scope of autonomous and rational decision making" (Meynen and Widdershoven 2013, p. 64). By questioning what they see as the capacity of individuals to incorporate danger and destruction in rational life-plans, Meynen and Widdershoven (2013), argue that the political conception of autonomy would unduly deny the right to seek physician assisted suicide, which has been established as a legal right in numerous jurisdictions, and conclude that "informed critical reflection should not a priori result in certain outcomes" (p. 65). Meynen and Widdershoven give the examples of extreme sports and Jehovah's witnesses refusal of blood donation as further instances in which my view allegedly leads to counter-intuitive results.

Again, I take this opportunity to clarify my argument, since it appears that the earlier formulation can be easily misunderstood. Namely, it is not the self-destructiveness that makes a life-plan unendorsable, but the tendency to either harm others directly, or lose self-control and possibly cause harm indirectly. A careful reading of my initial argument,[2] along with these clarificatory remarks should make it clear why extreme sports, physician assisted suicide and refusal of blood donation are not excluded (at least in principle) by the political conception of autonomy, but addiction to heroin in general is. Namely, practicing extreme sports or refusing blood transfusion may only physically harm the individual in question, not others. Also, while engaging in extreme sports or following the religious tenets of Jehovah's witnesses, it is quite possible to lead a life-style which does not produce indirect harm to others (by say stealing their property, or endangering their health), which is not really the case with heroin. Indeed, even if an extremely rich person (thus having no need to find meaningful work, or seek funds for purchasing drugs illegally) chooses to become a heroin addict, the issue of losing autonomy by depletion of self-control remains. An extreme sports enthusiast or Jehova's witness can effectively change their mind and stop with the activity (or belief regarding blood transfusion) in question without need for outside help and support, whereas the rich heroin addict cannot even in principle.

Now, let's turn to the more complex example of physician assisted suicide. Meynen and Widdershoven contend that the idea of a rational life-plan would curtail the

---

[2]"As a result of seeking access to drugs, heroin addicts often engage in risky, degrading, and illegal activities, which is not acceptable as a rational life-plan under fair terms of social cooperation. Also, they spend time and effort both in pursuit of their drugs and attempting to stop consuming the drugs. […] Heroin addicts oscillate between the preference to consume the drug and the preference to be free from the compulsion to use it because they are "ego depleted." That is, assessing rational options and inhibiting affordances and reflexive responses uses cognitive resources of self-control that their addiction depletes […]. Normally, these resources are sufficient so that healthy adult humans can plan their future actions reliably, without constant fear that they will impulsively do the opposite of the planned action" Dubljević (2013, p. 49).

autonomy of patients seeking death with dignity in principle. However, even though death means the end of autonomous decision making and is very much irreversible, for someone who is terminally ill and suffering debilitating amounts of pain, ending their life before their condition becomes worse and the pain becomes unbearable simply is a rational plan, in this case a rational end-of-life plan. The ethical issues regarding physician assisted suicide are not pertaining to the self-destructiveness, but to the potential danger of abuse of lax 'right to die' laws. Thus, on my view of autonomy, the state would never be justified in criminalizing suicide, but forcing *someone else* to commit suicide is legitimately prohibited and sanctioned. Participating in, or providing means for *someone else* to commit suicide was for a long time prohibited by default as well, but certain cases such as Sue Rodriguez in the United States and "Amy" in Canada[3] led to renewed scrutiny of these legal defaults. Public debates have led certain countries to allow the practice, which would mean that their exercise of public reason has resulted in the verdict of public autonomy that such rational 'end of life' plans and providing means are endorsable after a period of informed reflection and acceptable under fair terms of social cooperation. Now, the political conception of autonomy does not exclude the public autonomy of citizens in other jurisdictions who have concerns over pressuring the elderly into considering death prematurely without adequate medical reason or abuse in terms of euthanasia for persons experiencing cognitive decline. Thus, the 'informed critical reflection' is public in these cases, and in many others including legal sanctions.

These arguments should also put to rest Meynen and Widdershoven's concern that informed critical reflection a priori results in certain outcomes. This is far from being the case—in ordinary mundane matters, any person has the right to change their mind based on their implicit or explicit reasoning. In socially controversial matters, the reflection is part of public reason and as seen with the example of physician assisted suicide, public reason can produce very different outcomes. The fact that public reasoning on the dangers of heroin consistently produces agreement on heroin's tremendous harms for both individuals taking it and others (see e.g., Nutt et al. 2007,

[3]The case of "Amy" involves a seventy-seven-year-old patient who was by all accounts a bright and vivacious woman. After discovering that she had lymphoma, Amy had pretty blatantly expressed her wish to die. She made funeral arrangements for herself (including meeting the coroner in advance) and gave away her money and possessions. Her actions after her diagnosis did not appear to be sudden, spur of the moment decisions. To go through all of those steps, Amy obviously took time and careful consideration. She did her research and looked into other cases where people were seeking physician assisted suicide and during her research she came across the 1993 case of Sue Rodriguez. In that particular case, Sue Rodriguez petitioned the court but was denied the right to die with the assistance of others. Amy's candidness about suicide resulted in her being briefly admitted for suspected mental disease. Amy resented being subjected to psychiatric scrutiny and was assertive that this should be reserved for the mentally ill. She claimed that her own situation was different, since she was rational and had made a conscious, well-informed decision. As Dr. Cameron who brought this case to the attention of the academic public writes: "despite differences in opinion and considerable angst, the [hospital medical personnel] reached the consensus that Amy's interests would be best served by her discharge from hospital" (Cameron 1997, p. 231). A few days after the hospital discharge, the coroner, to whom she previously brought some chocolates in preparation for her death, informed Amy's family physician that Amy was dead, and that she drowned in the waters of the harbor.

who ranked it as the most dangerous drug) makes the assertions I made more or less uncontroversial.

6.9 Is Autonomy Both Political and Metaphysical?

Some objections concede the point of separate justifications for political concepts such as autonomy and metaphysical notions such as 'free will', but nevertheless assert that they are different aspects of the same phenomenon. Johannsen (2013), for instance, thinks that "independent justification doesn't entail that a different concept is under discussion" (p. 66). Johannsen also argues that the view of autonomy as a matter of degree raises some difficulties, which he believes are not insurmountable, and that determinism might make some entrenched political solutions, such as assigning normative significance to choice as opposed to circumstance, less workable.

I will start with the easiest issue: the difficulties Johannsen thinks are raised by a "degree view" of autonomy. Johannsen uses two examples, the link with personhood and the issue with equal protection, to make the point about difficulties. In this context, Johannsen asks two important questions "[i]f autonomy is to be thought of as a matter of degree, however, then does that also mean personhood is a matter of degree?" and "[i]f some people are capable of greater autonomy than others, however, then does that mean the protection of their autonomy matters more?" (Johannsen 2013, p. 66). The answer to these questions is a pretty straight-forward 'No' in both cases. To elaborate on the first point, Frankfurt's hierarchical account of autonomy is intended as an account of personhood, but the political conception of autonomy is not. Indeed, above I have noted that autonomy is legally recognized at certain ages, whereas minors are *persons* below that specific age recognized by law. It would be absurd to deny personhood to individuals who are 17 and grant it after year 18, whereas no such absurdity is entailed in the view of autonomy as a right. Similar holds for equal protection: some people exercise their autonomy more often than others, or are closer to the ideal of autonomy in terms of deliberating their decisions more, but the law does not (and should not) make exceptions even for exceptional people, and holds everyone to the same standard of the reasonable person.

Now, to consider the issue of different kinds of justification as raised by Johannsen. In a poignant passage, Johannsen concedes that the strategy of avoiding metaphysical disputes works well:

> Dubljević thinks that a conception of autonomy can be justified by virtue of its place in our political culture. If a conception of the autonomous citizen is embedded in democratic culture to the extent that it is generally accepted by members of a plural society, then […] it can justifiably be employed for purposes of ascribing responsibility in matters of law and morality […]. Since a justification of this sort doesn't seek to demonstrate truth so much as the permissibility of practical application, it is therefore also independent of metaphysical debate. An independent justification doesn't entail that a different concept is under discussion, however. (p. 66)

Johannsen is quick to add his worry about the metaphysical root of the concept of autonomy, whatever its political function may be. However, his worry rests on a mistaken assumption. Namely, he assumes that free will and autonomy are the same concept in different fields, justified with different means. However, this is not something he can assume without offering extensive evidence in terms of both empirical and historiographical data. Namely, my view starts from the premise that societies begin with the problem of 'permissibility of practical application' in terms of 'ascribing responsibility', and then some societies essentialize the solution of the problem in terms of metaphysical categories. As mentioned above in the examples regarding age, we have a strong intuition that children are not equal to adults in cognitive and volitional capacities required for responsible decision making, and we, based on social consensus and political and legal practice in democratic societies, deem adults autonomous and children non-autonomous. Similarly, in times past, people had a strong intuition that humans have some sort of property that animals do not. Given that these times were in the age of metaphysical thinking (Habermas 2004), this property was called a 'soul' or 'free will', and it neatly solved the problem of delineation of treatment for humans and animals. Animals could be owned and sold, exactly because they did not have a 'soul', whereas humans were thought to have this property. After a lot of wrangling about whether slavery might serve the purpose of saving the 'souls' of infidels (see Kitcher 2011, especially pp. 143–144), democratic societies settled the issue by forbidding any human being, and any organ from any human being from being bought and sold. My point is that practical application has precedence over metaphysical justification: if certain groups of people choose to associate autonomy with 'free will' or 'soul' based on their traditional religious or metaphysical beliefs, they are free to do so, but that does not mean that all others are forced to subscribe to their views. Indeed, to turn the tables on Johannsen—unless there is proof that autonomy cannot be justified by non-metaphysical means, there is no need to link it with 'free will', 'soul' or any other metaphysical notion.

Finally, Johannsen's worry about determinism making certain practices less justifiable, such as assigning normative significance to choice as opposed to circumstance, needs to be addressed. To this I answer first that neuroscientific findings point both to the stochastic (i.e., probabilistic) aspects of decision making and the deterministic aspects Johannsen fears would wreak havoc on certain political institutions. But even if determinism were scientifically proven to be true (as opposed to mere metaphysical assumption), I think this would not change our practices much. Consider the example of organ failure: [4] we consider it relevant if the cause of organ failure is precipitated by previous decision making of the person (e.g., heavy drinking) or something beyond the person's control (e.g., a genetic defect). But previous decision making simply is the best estimate of future decision making, whether the world is deterministic or stochastic. So a young child's chances of retaining a liver transplant are in many ways greater than a grown alcoholic's—whether the rationale for the decision to transplant the liver to the child is based on moralized norms of 'choice' versus 'circumstance' or on forward looking utility of the organ being donated, this

[4] I borrow this example from Kitcher (2011), pp. 379–380.

will make no difference in the final analysis. The bottom line is that current practices regarding 'choice' and 'circumstance' are not as controversial as the debate on determinism is.

6.10 Are Metaphysical Assumptions Required for Autonomy?

However, some critics might not be convinced by such arguments. They might contend that any use of the word 'choice' smacks of metaphysical commitments. Cameron Brewer and Jessica Morgan Gascoigne (2013) offer the sound advice to "avoid using language that implies some sort of free will" (p. 68). They also take aim at Rawls' political conception of justice and maintain, following Smilansky (2003), that any political conception of autonomy smuggles metaphysics 'through the window' after publicly throwing it 'out of the door'. Finally, they try to deconstruct my argument that free will is irrelevant for judgments of moral responsibility which leaned on a thorough review of multiple studies conducted by Roskies (2006).

Indeed, if the use of the word 'choice' is likely to throw metaphysically-minded readers onto assuming much more than is necessary, then by all means, I will strive to replace this vague term with 'decision making' in future writing. The fact that 'rational choice' has a long and distinguished history of use in science, does not change the issue: I have not used 'rational choice', merely the lay term 'choice', which might or might not have metaphysical baggage. This is more or less uncontroversial. The immediate problem is that Brewer and Gascoigne believe 'decision making' also implies a metaphysical position. To this I respond that to a person with a hammer everything looks like a nail—or in this case, to a metaphysician, everything looks like a metaphysical problem. 'Decision making' is a scientific term, denoting a phenomenon occurring in humans (see Edwards et al. 2007) and non-human animals (Conradt and Roper 2003), and is investigated with empirical methods of both natural and social sciences. As such, it is firmly on this side of physics or natural, empirically observable phenomena, and not part of metaphysics as an abstract concept. Thus, I consider it sufficient to heed part of their advice and use 'decision making' while avoiding 'choosing'. In this sense the definition of autonomous actions I provided (in Dubljević 2013 and Chap. 3) needs to be amended to the following:

> Autonomous actions could be analyzed in terms of competent *decision makers* who act (1) voluntarily or intentionally (volitional component), (2) with sufficient information and understanding (cognitive component), and (3) without controlling influences that would override previously rationally made *decisions* (liberty component).

Now, since the 'devil is in the details' I can only thank Brewer and Gascoigne for the advice to mind my language and avoid terms that are potentially loaded with metaphysical baggage. So far so good. How about the other two worries they raise? Brewer and Gascogne claim that "neither Rawls nor Roskies lend the necessary support to Dubljević's argument" (p. 68). In response, I should note that my leaning on Rawls and Roskies is of two kinds: to Rawls I credit the *conceptual* insight that moral-political concepts are independent from metaphysics and to Roskies I credit a careful review and analysis of *empirical* data. So, when I make the claim that 'free will of any kind is irrelevant for judgments on moral responsibility' I am referring to the data, which finds that people surveyed always think that others should be held accountable, whatever their judgments of determinism and free action might be. Now, Brewer and Gascoigne single out only one study done by Nichols and Knobe (2007), in which they are engaged in a "negative program" of experimental philosophy.

Namely, there are two conflicting programs of experimental philosophy in the study of moral intuitions (Sytsma and Livengood 2016): the positive and the negative program. The positive assumes that humans have a faculty of intuition which is fallible, but nevertheless tracks the truth of certain propositions, and uses survey sampling and statistics in order to aggregate intuitions to get a better glimpse of the truth. The negative program assumes that faculty of intuition is unreliable or insensitive to sources of bias, and aims to reduce 'folk intuitions' to holding absurd or inconsistent views (see Sytsma and Livengood 2016, p. 31). Thus, when Brewer and Gascoigne ask "[w]hy should we allow inconsistent folk intuitions to guide our notions of freedom and responsibility?" (p. 68), they are at the same time proclaiming their allegiance to the negative program and making perfectly clear why they have cherry-picked only one study for a specific *interpretation of data*. Now, we might disagree which program (positive or negative) is ultimately right, but this is an *empirical issue*: more studies will ultimately show if our intuitions about moral responsibility can be trusted or not, and in which circumstances. I suspect that for real-world scenarios they will be found to offer sound guidance. In the meantime, different interpretations of data are allowed, as long as they are faithfully reporting the data, and not relying on only one study while disregarding the others. Thus, on one plausible positive program reading (to which I subscribe), people always already ascribe moral responsibility, whereas intuitions about determinism are secondary, and are not able to subvert the more fundamental moral insights.

This fits very nicely with the explanation that moral-political concepts are separate and independent from metaphysical notions. And this is related to the point about Rawls that Brewer and Gascoigne have been making. Namely, they closely follow Smilansky (2003) in saying "People are expected to fulfill their obligations, to take responsibility for their actions [...] that makes sense only if non-hard determinist position on free will problem are assumed" (Brewer and Gascoigne 2013, p. 68). However, their challenge begs the question since (following Smilansky) Brewer and Gascoigne assume that a non-hard determinist position on 'free will' is necessary for moral responsibility, whereas, I (following Rawls) assume it is not. It is possible to settle the issue by recourse to intuitions of those untainted by philosophical 'indoctrination': members of the public. And this is exactly what the empirical studies reviewed in Roskies did—in Sytsma and Livengood's words they were "testing whether the intuitions reported by philosophers about various cases found in the philosophical literature are in line with the intuitions of non-philosophers" (2016, p. 34). The data can be interpreted in different ways of course, but looking at a range of studies and not only one, it is clear that philosophical intuitions about the necessity of either denying determinism or denying freedom are not shared by the general public—they first and foremost assume moral responsibility. This means that it is merely a social fact that people mutually expect to fulfill obligations and to be responsible for their behavior. Further metaphysical theorizing was done in ages past in order to make sense of that fact, and it has come up with some ingenious (if controversial) solutions. Now that we have no need to reconcile theology of an 'omnipotent God' who ordains events and moral responsibility, we can use the fruits of "postmetaphysical thinking" (Habermas 2004) to study the moral-political phenomena conceptually and empirically. This is not to say that metaphysics is irrelevant or not allowed—only not necessary or assumed in the discussion of moral-political concepts such as autonomy.

6.11 Is Autonomy as a 'Negotiated Concept' Merely Compromise?

Dolan (2013) raises the worry that the notion of autonomy will continue to shift, and that this shift cannot be predicted, but ultimately depends on the political discourse. He gives the example of the United States, in which reactions against 'the nanny state' asserted the "liberty" component seemingly over the volitional and cognitive elements.

Dolan starts his objection by first supporting the political conception of autonomy, and arguing that autonomy and informed consent are related, socially negotiated achievements, which are correlated in their close matching of the three conditions (volitional, cognitive and liberty). Dolan emphasizes that many historical documents (Nuremberg Code, Declarations of Helsinki, World Health Organization International Ethical Guidelines, Belmont Report and United Nations Guidelines on Free, Prior and Informed Consent) follow the three elements, and expand the purview of autonomy on an ever increasing scope. In Dolan's words "informed consent, a core practice in human subjects research that assumes autonomy, has been reconceived over time through a political deliberative process between other values and interests ranging from the moral to the pragmatic" (p. 70).

First of all, I need to emphasize the fact that autonomy as a value in democracies in not merely a product of political compromise, but an essential part of moral-political consensus. Dolan is right to point out that political discourse in democratic societies shapes how certain important practices, such as informed consent, are implemented. However, the political concept of autonomy actually logically constrains the practice of democracy and the shift can be assumed to be toward more inclusion. Namely, the practice of democracy assumed that people are equal in the sense of having the default volitional, cognitive and liberty components: all adults are assumed to have sufficient power of self-control, mental faculties and are sufficiently independent of undue coercion or compulsion. The key issue was to recognize that assigning and recognizing autonomy needs to go beyond the enfranchised citizens to include women (for a long time suspected of lacking mental faculties), racial minorities (for a long time accused of lacking self control), people of different religious/metaphysical, political or sexual orientations (for a long time suspected of being unduly coerced or compelled) and even non-citizens or stateless persons. The shift is not so much a deliberative process of weighing different interests, but a logical expansion of a fundamental principle to include more and more individuals and groups and offer them normative protection for their legitimate decision making. So, is the liberty component more entrenched over the cognitive and volitional components? Not necessarily—the need to limit state coercion is ever present in all democracies, but it is the cases of loss of cognitive faculties (such as in Alzheimer's) and loss of voluntary self-control (such as in Frontotemporal dementia and other disorders) that might be pushing and probing the limits of autonomy in the age of increasingly aging populations in democracies.

Then again, Dolan might be correct in asserting that "the information revolution might transform the negotiation under the mantle of transparency", effectively "bringing a new impetus to making individual/collective decisions comprehensible" (Dolan 2013, p. 70). Time will tell, but autonomy is sure to stay a central issue in democratic societies.

6.12 Does a Rawlsian 'Original Position' Presuppose Metaphysics?

Finally, some objections start from a conviction that the political constructivist mode of thinking presupposes metaphysical knowledge. More specifically, Simkulet (2013) argues that the distinction between autonomy and free will "is a mistake, [...][since] our political conception of autonomy has clear and practical metaphysical underpinnings relevant to participants in Rawls's original position" (p. 71).

Drawing on Galen Strawson, Simkulet claims that the presence or absence of feeling of freedom dictates when it is appropriate to hold people accountable for their actions. Simkulet gives the example of impulse-control disorders to argue that on the one hand this feeling of freedom is lacking, and on the other hand social approbation shifts in the recognition of the compulsive nature of certain actions.

However, the example Simkulet gives in no way challenges the political conception of autonomy. We always already presuppose that healthy adults are responsible for their actions. If there is clear evidence of pathology, we do not casually disregard such asocial actions, but respond by denying the rights of autonomy, and treating the affected individuals paternalistically, akin to children. We still strive to prevent anti-social action, but assign other competent adults as their legal guardians. But what about Simkulet's argument about the 'original position'? In a lengthy passage, which needs to be considered in full in order to appreciate the argument, Simkulet contends that:

> [i]n the original position, participants are tasked with determining the rules that they would choose to live under in a society. Stipulated to be rationally self-interested, participants would be inclined to favor rules that would benefit themselves, and unanimous agreement would be unlikely except for the veil of ignorance, which keeps participants ignorant about contingent facts about themselves—their age, race, sex, religion, skill set, and so on. Although participants know that people of different ages, races, sexes, and so on exist, they would not agree to put any group at a disadvantage because they might actually be a member of that group.
>
> Now suppose you found yourself in the original position, under the veil, and were tasked to construct principles concerning how to deal with those who violate the principles accepted in the original position. Here it is clear that it would be instrumentally valuable to know whether our actions are wholly causally determined, or whether our experience of authorial control is genuine. (Simkulet 2013, p. 72)

Now, even though technically Simkulet is right that such knowledge could theoretically be beneficial, the certainty for it is completely lacking. Namely, an entirely different reading of what information on determinism is available in the original position is more plausible: that there are conflicting metaphysical positions of hard determinism, libertarianism and compatibilism, and that participants in the original position are ignorant of contingent facts of their own metaphysical commitments. Participants know that people of different metaphysical bents exist, and they would not agree to put any group at a disadvantage because they might actually be a member of that group. Therefore, they opt for the principles such as justice and autonomy to be enshrined in law, so that they do not favor any particular group, and which could be well implemented within any reasonable comprehensive doctrine they might end up endorsing.

References

Adeleye, O.A., and E.A. Tobin. 2013. Do hospitalized patients in a Nigerian community consider informed consent necessary? *AJOB Primary Research* 4 (2): 51–56.

Bell, E., V. Dubljević, and E. Racine. 2013. Nudging without ethical fudging: Clarifying physician obligations to avoid ethical compromise. *American Journal of Bioethics* 13 (6): 18–19.

Brewer, C.D., and J. Morgan Gascoigne. 2013. The metaphysical assumptions required for political autonomy. *AJOB Neuroscience* 4 (4): 67–69.

Bryan, H., and V. Dubljević. 2017. To disclose or not to disclose: When fear of nocebo effects infringes upon autonomy. *American Journal of Bioethics* 17 (6): 50–52.

Cameron, S. 1997. Learning from Amy: A remarkable patient provokes anguished debate about rationality, autonomy and the right to die. *Canadian Medical Association Journal* 156 (2): 229–231.

Conradt, L., and L. Roper. 2003. Group decision-making in animals. *Nature* 421 (6919): 155–158.

Dewey, J. 1921/2002. *Human nature and conduct*. New York, NY: Dover.

Dolan, T. 2013. Autonomy as a negotiated concept: The case of informed consent. *AJOB Neuroscience* 4 (4): 69–70.

Dubljević, V. 2013. Autonomy in neuroethics: Political and not metaphysical. *American Journal of Bioethics – Neuroscience* 4 (4): 44–51.

Dubljević, V. 2015. Lost in interpretation: Autonomy and what patients tell versus what is inferred. *The American Journal of Bioethics* 15 (9): 28–30.

Dubljević, V. 2016. Autonomy is political, pragmatic and post-metaphysical: A reply to open peer commentaries on 'autonomy in neuroethics'. *AJOB – Neuroscience* 7 (4): W1–W3. https://doi.org/10.1080/21507740.2016.1244125.

Edwards, W., R.E. Miles Jr., and D. von Winterfeldt (eds.). 2007. *Advances in decision analysis: From foundations to applications*. New York: Cambridge University Press.

Felsen, G., and P.B. Reiner. 2011. How the neuroscience of decision making informs our conception of autonomy. *American Journal of Bioethics - Neuroscience* 2 (3): 3–14.

Felsen, G., and P.B. Reiner. 2013. Having the capacity for autonomy is insufficient to provide meaningful autonomy. *AJOB Neuroscience* 4 (4): 52–53.

Frankfurt, H. 1971. Freedom of the will and the concept of a person. *The Journal of Philosophy* 68 (1): 5–20.

Frankfurt, H. 1988. *The importance of what we care about*. Cambridge: Cambridge University Press.

Greene, J., and J. Cohen. 2004. For the law, neuroscience changes everything and nothing. *Philosophical Transactions of the Royal Society in London* 395: 1775–1785.

Habermas, J. 2004. *Between facts and norms: Contributions to a discourse theory of law and democracy*. Cambridge: Polity Press.

Horstkotter, D., and A. Snoek. 2013. The importance of the self for autonomous behavior. *AJOB Neuroscience* 4 (4): 62–63.

Hui, K. 2013. The illusion of post hoc autonomy. *AJOB Neuroscience* 4 (4): 57–58.

Johannsen, K. 2013. Free will and determinism: Political, not just metaphysical. *AJOB Neuroscience* 4 (4): 65–67.

Kitcher, P. 2011. *The ethical project*. Cambridge, MA: Harvard University Press.

Lettieri, D.J., M. Sayers, and H.W. Pearson (eds.). 1980. *Theories on drug abuse, National Institutes of Drug Abuse Research Monograph 30*. Washington, D.C.: U.S. Government Printing Office.

Levy, N. 2013. Free will doesn't come for free. *AJOB Neuroscience* 4 (4): 53–54.

Meynen, G., and G. Widdershoven. 2013. Autonomy, free will, and a rational life-plan: A practical perspective. *AJOB Neuroscience* 4 (4): 64–65.

Morse, S. 2007. The non-problem of free will in forensic psychiatry and psychology. *Behavioral Sciences & the Law* 25: 203–220.

Nagel, S.K. 2013. Autonomy—A genuinely gradual phenomenon. *AJOB Neuroscience* 4 (4): 60–61.

Nichols, S., and J. Knobe. 2007. Moral responsibility and determinism: The cognitive science of folk intuitions. *Nous* 41 (4): 663–685.

Nutt, D., L.A. King, W. Saulsbury, and C. Blakemore. 2007. Development of a rational scale to assess the harm of drugs of potential misuse. *The Lancet* 369 (9566): 1047–1053.

Pinker, S. 2011. *The better angels of our nature*. New York: Penguin.

Rawls, J. 2005. *Political liberalism*, Expanded ed. New York: Columbia University Press.

Roskies, A. 2006. Neuroscientific challenges to free will and respon- sibility. *Trends in Cognitive Science*. https://doi.org/10.1016/j.tics.2006.07.01.

Sharp, D., and D. Wasserman. 2013. Compatibilism and a political conception of autonomy. *AJOB Neuroscience* 4 (4): 55–56.

Simkulet, W. 2013. Autonomy as free will. *AJOB Neuroscience* 4 (4): 71–72.

Smilansky, S. 2003. Free will, egalitarianism and Rawls. *Philosophia* 31 (1–2): 127–138.

Sytsma, J., and J. Livengood. 2016. *The theory and practice of experimental philosophy*. Peterborough, ON: Broadview Press.

van der Eijk, Y. 2013. A blurry line between metaphysical free will and autonomy in addiction. *AJOB Neuroscience* 4 (4): 58–60.

Wegner, D.M. 2002. *The illusion of conscious will*. Cambridge: MIT Press.

Chapter 7
Are There Problems with the Economic Disincentives Model of Regulation?

In the chapter on empirical constraints of psychopharmacological cognitive enhancers (Chap. 4), I analyzed available information and policy options for the two of the most commonly used cognitive enhancement (CE) drugs: Adderall and Ritalin. I concluded that for all forms of amphetamine, including Adderall, and for instant-release forms of methylphenidate, any form of sale beyond prescription for therapeutic purposes needs to be prohibited, while some form of a taxation approach (see Dubljević 2012a) and the economic disincentives model (EDM) in particular (see Chap. 1 and Dubljević 2012b) could be an option for public policy on extended-release forms of methylphenidate (such as Ritalin-SR).

However, there has been a considerable amount of criticism regarding my proposal for regulation of CE drugs. Some neuroethicists objected to my favoring prohibitive policies to dangerous CE drugs such as amphetamine and argued for laissez-faire or even mandatory use of enhancements. Others took issue with the conclusion that the economic disincentives model (EDM) could be an option for public policy on extended release forms of methylphenidate. Furthermore, there are those that think my argument in general and EDM in particular are failing to address the relevant issues in regulation of CE, such as social justice and real autonomy. Finally, there are those who offer suggestions on how the argument and the model of public policy for CE drugs can be improved.

7.1 The Value of Proper Information on CE

Sandberg (2013) insists that there could exist other enhancers besides classical stimulants (he gives the example of modafinil), enhancers that might have very different

This Chapter draws and expands on my previous work which has been published as: Dubljević (2014).

V. Dubljević, *Neuroethics, Justice and Autonomy: Public Reason in the Cognitive Enhancement Debate*, The International Library of Ethics, Law and Technology 19,
https://doi.org/10.1007/978-3-030-13643-7_7

usage and risk profiles and that are likely to lack autonomy-impairing addiction, withdrawal and tolerance properties. According to him, they are likely to be even safer than the extended-release methylphenidate and are not legislated by international treaties, such as the 1971 United Nations (UN) convention on psychotropic drugs (UN 1971). As such, they should be regulated even more permissively. Sandberg contends that the greatest problem for regulation is insufficient information, and that the method used to regulate enhancers will affect what information will become available for fine-tuning policy.

Although we both agree on the point that bans and laissez-faire approaches do not provide the necessary information, Sandberg argues that adding taxes and fees as suggested in EDM does not give much feedback except perhaps providing usage statistics. Even though he acknowledges that having licensed users undergo regular medical tests (as envisioned by EDM) would provide more relevant information, he considers that the EDM would risk creating principal-agent problems between the interests of users, companies, testing bodies, and society in general. Society would benefit from extensive and careful testing, while users would not be interested in sacrificing too much money and privacy for it. Sandberg concludes that the way to minimize harm would be to accumulate relevant information as early and accurately as possible, which entails a liberal permissible regulation of safer CEs.

I heartily agree with Sandberg that regulatory models that could provide the missing information would be more effective, even if their preliminary assumptions turn out to be incorrect in the long run. I also agree that there might be more CE drugs to which a moderately liberal regulatory approach could be applied. For example, EDM could and perhaps should be applied to modafinil (see Dubljević 2016a for a longer argument)[1] due to the specific social risks it entails. However, even though Sandberg seems to disfavor EDM as a regulatory option, he failed to provide a realistic alternative model, so, apart from extending EDM, there is no feasible alternative option from which to choose. Even though EDM might need lots of fine tuning, that can only be done once it (or a similar model) is implemented and the information on the health costs associated with CE drug use becomes known.

Sandberg's other concern is with the acceptability of the licensing procedure and other measures in EDM. However, similar requirements are accepted worldwide in the case of vehicles: in order to use them, a person must pay fees for a training course and pass an exam as proof of competence. Then, when the vehicle is bought, taxes should be paid. In order to use the vehicle, an appropriate insurance must be taken and both the vehicle and the driver should be registered by a government agency. Finally, while using the vehicle, taxes on fuel, tolls, and appropriate fees for regular technical checkups must be paid. Since all these measures are readily accepted, there is no reason to doubt the acceptability of similar measures in EDM. But others have

---

[1]The key reason modafinil is not discussed in this book is that a crucial element for evaluating harmfulness—thorough assessment by means of the Multi-Criteria Drug Harm Scale is still missing (see Dubljevic 2018). Namely, societal costs of possible injury to health through neuroenhancers is adequately captured by the use of the Multi-Criteria Drug Harm Scale, and this information is available for 'classical stimulants', but not newer classes of drugs. Future work will have to address this gap in research.

different objections: CE can be achieved in more ways than one. Instead of medical drugs, people might be using trace elements, vaccines, or medical devices.

7.2 Mandatory Use and Other Forms of Enhancement

Levy (2013) agreed that EDM may be appropriate for regulating extended-release forms of methylphenidate and that amphetamines need to be prohibited. After this endorsement, he took issue with my ruling out of mandatory use for other forms of CE. He argued that in other instances, the costs can be negligible or low enough, and the benefits great enough, to make it appropriate to require enhancement. He gives the examples of vaccination and fluoride in the water supply, which allegedly override any informed choice regarding use. Furthermore, he claims that medical devices that can be used in the transcranial direct current stimulation (tDCS) modality are effective as enhancers and risk free.

As for Levy's comments about the applicability of mandatory cognitive enhancement use and his example of fluoride, there is a huge difference in providing trace elements that might increase the health of the population and providing mind-altering drugs. That being said, the use of fluoride in his example is actually not a case of mandatory use, but merely 'encouraged use' by the government. Many people purchase bottled water or use filtering such as active carbon or reverse osmosis, and the coercive power of the state is not brought to bear on them to change these practices. Mandatory use is indeed very hard to justify. Even vaccination is mandatory mostly for minors—in this case the state could sometimes override the wishes of the parents because the life and interests of a person not yet capable to make autonomous choices is at stake, and the parent's choice could be dictated by religious or other reasons that the minor might not endorse, or might choose to abandon after reaching maturity. However, in the case of adults, barring a major health disaster (e.g., a plague outbreak), coercion backed by the immense power of the state needs to be limited, even if society and individuals might perhaps benefit by state intrusion.

To give one example, governments that have historically made mind-altering drugs mandatory—for example, Togo militarists in Japan during the Second World War (for more details see Iversen 2008)—have found it easy to make such legal requirements. Making the use of a specific substance mandatory is simple: a law that suits the purposes of the elite is simply enforced and backed by sanctions. However, such a law would not be legitimate or democratic. Levy is right in that I have dismissed mandatory use without too much space allocated to discussing the issue. However, I have relied on democratic values and the overwhelming demand to respect the autonomy of citizens as being assumed (and discussed as relevant in Chap. 3 of this book).

Finally, Levy mentions tDCS devices as safe and effective cognitive enhancers. However, apart from the problems with requiring use of any sort of enhancement, the assertion about safety and efficacy of tDCS needs to be taken with caution. Here I'd like to thank Levy for prompting me to undertake the analyses that form the basis of Chap. 5 in this book. To remind the reader of the conclusions from that discussion—although investigative use of tDCS does appear safe and effective from a strictly scientific standpoint (e.g., in controlled laboratory settings), enhancement use without supervision might cause serious adverse effects, such as temporary respiratory paralysis (Brunoni et al. 2011). In addition, even in scientific studies, potentially detrimental effects are not measured and hence not reported, even though tDCS enhances certain cognitive functions while inhibiting others (Iuculcano and Cohen Kadosh 2013). Although for reasons of space I cannot go into more detail here (and must refer the reader to Chap. 5), this should suffice to prove that a regulatory approach that enforces being trained in the use of CE and provides information on detrimental effects should take precedence over pro-enhancement regulatory proposals.

However, some might be unconvinced that my conclusions regarding autonomy in the case of stimulant use are sound, and presume that a liberal-consequentialist perspective would render any prohibitive response illegitimate.

7.3 Conservatism and Liberal Consequentialism

Savulescu (2013) alleges that my proposal is too conservative and too prohibitive when viewed from a liberal consequentialist perspective. He relies on the argument from Mill's *On Liberty* (1859) to show that any intrusion of the state is illiberal and undemocratic. He takes issue with my claim that the use of amphetamines (like Adderall) would lead to the undermining of autonomy and cause addiction, and argues that addiction is primarily imprudent pleasure seeking and that addicts are not incompetent and need not be harming anyone else. Consequently, they should be free to harm themselves, and the state should not interfere with their freedom.

First of all, Savulescu's claim that my proposal is too conservative is problematic. Since "conservative" is defined as "averse to change or innovation and holding traditional values" (Oxford Dictionary 2013) and the EDM proposes a drastic change in regulation of stimulants such as methylphenidate and non-invasive brain stimulation devices such as tDCS, my proposal is everything but conservative. He might be opposed to prohibition of amphetamines, but since amphetamines can cause aggression, impulsivity, manic behavior, and psychotic episodes, and so can cause considerable danger to users and others, a form of prohibition might be legitimate even based on a liberal-consequentialist reading. However, one point about the kind of prohibition should be noted here: namely, in Chaps. 3 and 4, I emphasized that if an individual, for whatever reason, voluntarily and autonomously chooses to consume illicit drugs with full knowledge of their addictive properties and harmful physiological and social consequences, the society would be legitimate in punishing the producers and distributors of illicit drugs, while drug addicts might need to be treated and not punished. To conclude, an argument based on autonomy does exclude certain types of prohibition (i.e., prohibition of possession and use) but not others (i.e., prohibition of production and sale). However, precisely the issue of prohibition elicited other points of view. Not everyone thinks that my proposal is too conservative. Quite to the contrary, some object to EDM as too permissive.

7.4 Putative Constraints for Regulatory Options

Hall et al. (2013) claim that my proposal for EDM sounds plausible but that there are a number of major problems with it. They insist that it should not be assumed that sustained-release forms of drugs will be safer than immediate-release forms when these drugs are widely used in the community. They also have doubts about the feasibility and effectiveness of EDM. They think that tight regulations, licensing, and high taxes would be major disincentives for both the pharmaceutical industry and would-be stimulant users, so EDM may actually boost the gray or black market for stimulants.

Hall et al. (2013) also contend that EDM would require a modification of the 1971 UN convention on psychotropic drugs (UN 1971). Furthermore, they think that the bioethical debate about enhancement use of stimulants might have disastrous consequences: Proposals for liberalization of enhancement use of stimulants might actually lead to severe restrictions if not prohibition even of medical use of these substances, because society might recognize such use as an increasing social problem. This would make the population that really needs these medications as treatment for attention deficit hyperactivity disorder (ADHD) and other conditions vulnerable.

I agree with Hall and colleagues that it should not be taken for granted that sustained-release forms of drugs will be safer than immediate release forms. Indeed, in my proposal I have insisted that the danger profile of methylphenidate would have to be carefully analyzed and empirical studies confirmed by independent researchers before any change in current prohibitive policy is allowed. As for the worry that EDM may actually boost the gray or black market for stimulants, it is more or less uncontroversial to assume that EDM or any other taxation approach policy would not boost it any more than prohibitive policies do. On the contrary, if there are legal means of marketing or obtaining a commodity, the majority will prefer to act within the bounds of the law, thus avoiding sanctions and reputational costs.

Regarding the conclusion that EDM, or any other liberal policy, would require a modification of the 1971 UN convention, their argument is less than convincing. Article 3 of that convention states that a preparation may be exempted from the current regulatory regime if it is compounded in such a way that it presents no, or a negligible, risk of abuse and the substance cannot be recovered by readily applicable means. As I noted earlier, whether this is the case with extended-release forms of methylphenidate is an empirical question that can be settled with sufficient research, and this research needs to be conducted by independent laboratories, since the pharma industry has a vested interest in loosening of the drug control regime that has brought huge socio-medical costs to society on more than one occasion (e.g., the opioid crisis in the U.S.).

Finally, adverse reactions by conservative political factions are hardly reason enough to censor the freedom of bioethical and neuroethics discussion. Indeed, different policies are (and should be) judged by their merits—including the ability to provide reliable information on the prevalence of use. Therefore, even if there were a rash prohibitive response to liberal policies on stimulants that would restrict medical uses, such responses would be short-lived, as they benefit no one and harm the interests of many citizens.

7.5 On the Indirect Psychological Costs of CE

Faulmüller et al. (2013) argue that new cognitive enhancers are psychologically different from other well-known drugs such as caffeine, because they are perceived negatively by the public. They point out that psychological valence accounts for much, as some of the "old" substances like alcohol might be objectively more dangerous than methylphenidate. However, they are loosely regulated because they are perceived more favorably, and even as socially desirable. Moreover, not only are new enhancers judged negatively, but their efficacy is strongly exaggerated by lay people. Independently of any direct negative physiological effects, enhancers might generate some indirect psychological costs, such as attribution of performance (any success users might achieve would be attributed to the enhancer), dehumanization (users might be perceived as being more similar to automatons), and ostracism (enhancement users may be shunned by others).

I agree with Faulmueller and colleagues that views of the lay public need to be taken into account. However, these views can also be changed as a result of sufficient information. Consider yet again the example of tobacco: Smokers used to be considered socially more apt and even desirable (e.g., as portrayed by the character of Jim Stark (played by James Dean) in *Rebel Without a Cause*), but thanks to the information on objective harms of tobacco, nowadays smoking is more likely to be seen as a sign of weakness or poor taste. Indeed, smokers are increasingly ostracized, and that is precisely the point of any "discourage use" policy, including EDM. As for dehumanization and misattribution of performance, even though some people might have such exaggerated reactions to the use of enhancers, these concerns are generally matters relevant for individual choice, but not relevant for public policy (see e.g., Dubljević 2012a). Legitimate public policies in democratic societies usually protect autonomous choice of individuals, as long as this choice does not harm others (see Chap. 3, and the discussion in Chap. 6). However, some might be convinced that the analogy with tobacco I used in Chap. 4 is precisely the reason to support a more prohibitive response.

7.6 On the Analogy Between CE and Tobacco

van der Eijk (2013) claims that arguing for EDM as a policy for CE drug use based on an analogy with tobacco might not be persuasive, since tobacco control policy is actually taking a turn toward prohibition. She explains the idea of a "tobacco endgame": the complete phasing out of tobacco consumption by prohibiting sale to newer generations even when they reach the age of adulthood. Apparently, this idea has become popular in the tobacco control literature in the last decade, and is being considered in several countries. She defends such prohibitive responses by pointing out the addictive properties of nicotine, and further distinguishing between recreational and "addiction maintenance" use. The addictive properties of nicotine guarantee a high proportion of "maintenance smokers", which in turn means that discouragement policies are unlikely to have much effect. Hence, the only really feasible solution would be to prevent people from becoming addicted in the first place by prohibiting the substance for the newer generation and "phasing out" the "old" users.

I agree with van der Eijk that there is sufficient evidence, provided by the Multi-Criteria Drug Harm scale assessment, on which I draw in my analysis, that smoking is a nasty habit that causes considerable amount of physiological and social harm and often leads to dependence (see, e.g., Nutt et al. 2007). Indeed, most long-term smokers need substantial interference by outside factors in order to get rid of the habit, and even if they are successful there is the danger of relapse (as an ex-smoker I am painfully aware of this). However, we should keep in mind that protecting autonomy and life plans of some should not be done by severely restricting autonomy and life plans of all, especially if there is clear evidence that individuals are able to stop the habit if they so choose (and I would be one example of that—see also discussion below). A ban on smoking in public places suffices to protect the legitimate interests of other and a harm that is restricted to self should not be made illegal lightly; that would mean that, for example, a "tattoo-free" generation might be the next move of some conservative factions.

Van der Eijk's argument on "addiction maintenance" has more merit. However, not every dependence is the same. In Chap. 3, I provide a longer argument, but in this context I think it suffices to note that dependence on nicotine does not entirely disrupt all other rational life plans a person might have, whereas addiction to, say, heroin does. For instance, the reader should remember that while smoking, a person can also be meaningfully socially connected in various capacities. Opposed to that, full-blown heroin addicts will often engage in risky, degrading, and illegal activities, which is not acceptable as a rational life plan under fair terms of social cooperation. If smokers want to keep their habit in the privacy of their homes, and observe the ban on smoking in public places, a liberal democratic society would be hard pressed to find faults in their right to do so.

The question of arbitrary discrimination against younger adults is an additional reason to believe that "end games" or any other prohibitive regulation of tobacco would be a failed and illegitimate social policy. Indeed, even though the idea is being considered in several countries, it is more or less uncontroversial to assert that such a response would be short-lived, at least in liberal-democratic countries.

7.7 On the Ethics of a Better "Smart Drug"

LaBuzetta (2013) agrees that the case for EDM as legitimate regulation of Ritalin and Adderall in a democratic society is compelling. However, she thinks that it can nevertheless be problematic when the same rationale is applied to other drugs with improved risk to benefit ratios. When applied to safer "smart drugs" such as modafinil, an evidence-based regulatory model might actually seem to compel their use. La Buzzeta admits that such a compulsion would be contrary to the very idea of liberty and autonomy, but thinks that perhaps there are other ideas and values that should take precedence.

She argues that it is plausible and reasonable to demand that individuals working in high-responsibility roles, such as military, medical, and aeronautical professions, use a safe and effective cognitive enhancing medication, such as modafinil. She considers different principles that could support her conclusion and opts for utilitarianism.

First of all, I am well aware that Chap. 4 of this book left out some very important CE drugs, such as modafinil, but this was due to reasons of space, and the lack of adequate information based on the Multi-Criteria Drug Harm Scale (see Dubljević 2018). However, I believe that EDM can be successfully extended to most new CE drugs that are not too dangerous and/or addictive, and I have argued elsewhere (Dubljević 2016b) that this is the case with modafinil. The reason for this is that newer CE drugs are likely to have many unknown effects. Even if clinical studies prove they are safe and effective for specific pathologies, the data on effects of prolonged use need to be somehow generated. EDM with the requirement of annual medical testing and additional insurance is specifically designed to fill that gap.

Regarding mandatory use by certain professions such as pilots, LaBuzetta is right to note that these are issues that EDM did not tackle. Indeed, military pilots are frequently ordered to take even dangerous amphetamines on prolonged combat missions. This practice might be seen as wrong, or as justified by the inherent danger of combat missions. Whatever our take on this specific practice, issues of autonomy and responsibility need to be taken into account. For example, legal representatives of U.S. Air Force pilots who have killed Canadian soldiers in a "friendly fire" incident while under the influence of Dexedrine (dextro-amphetamine) argued that amphetamine use has diminished autonomy and responsibility of their clients (see Bigelow 2006, 238). That is one of the reasons why I think that La Buzzeta's choice for utilitarianism as the right ethical framework for regulating these specific cases is problematic. In addition to that, it demands too much information and cognitive resources for decision making, which is not available and thus is not feasible.

Even though I do not have the space to engage in a protracted discussion on Utilitarianism, it is fairly safe to assert that this ethical theory requires that all alternative responses are considered, to calculate all consequences of all identified options, to predict and develop contingency plans for all unintended consequences, and to calculate the probability that a certain response sets a precedent for other circumstances where the information may be less reliable. All this creates a nice optimization model, but no mind or machine can solve moral problems in this way. In the real world, even with "safer" smart drugs we just don't know enough to rely on foreseeable consequences alone. That is why in any discussion on regulatory options, in addition to consequences, we need to take rights and virtues seriously. However, others have different objections: Even if we opt for deontological principles, such as justice, EDM may fail to promote it, despite its initial conceptual impetus.

7.8 On EDM and Social Inequality

Brewer and De Grote (2013) argue that if principles of biomedical ethics (Beauchamp and Childress 2009) are considered more thoroughly (specifically, the principles of justice and nonmaleficence), EDM for the regulation of stimulant drug use must be rejected. They claim that EDM is at odds with the "fair opportunity" rule: It would deny the social benefits of enhancement on the basis of undeserved disadvantageous properties. They grant that some form of "discourage use" model might be applicable, but that it should not be as costly to the consumer as EDM.

According to Brewer and De Grote, EDM would almost certainly create an even greater gap between the haves and have-nots. It would effectively deny the social benefits of extended-release formulas of methylphenidate to the economically disadvantaged. Furthermore, they argue that EDM would harm the interests of the disadvantaged, as it would exacerbate the already increasing academic achievement gap between the rich and the poor. EDM as the regulatory policy for enhancement uses of stimulant drugs would ensure that only those with higher incomes would be able to buy and use methylphenidate, while the poor would have to keep their uphill struggle for academic achievement without it.

Since EDM has been developed precisely as a means to uphold justice and protect the rights and interests of the disadvantaged (see Chap. 1), this is a serious objection. However, Brewer and De Grote's argument would only be convincing if EDM would make CE drugs so expensive as to be unaffordable for the poor. Consider once more the analogy with tobacco: Discourage-use policies on tobacco introduce taxes, which are sometimes very high. In Norway, they amount to 200%. And yet, even though cigarettes are not cheap, most smokers in Norway live in poorer regions and earn low incomes. Now, it is obvious that extra taxation did not make tobacco unaffordable to the poor. Indeed, EDM envisions that the prices of CE drugs would be regulated and that the policy would be fine-tuned based on the feedback information on the use patterns. This means that regulation of prices could be relaxed if it turns out that the poor are disadvantaged in academic achievement due to the price of CE drugs.

An additional problem for Brewer and De Grote's objection is that they seem to assume that CE drugs would offer only a competitive advantage, and no health disadvantages (for a longer argument, see Chap. 2). This is problematic in several respects. For example, even though nicotine can also be seen as a mild cognitive enhancer, it offers mostly health disadvantages. Given that stimulants are known to cause nervousness, drowsiness, insomnia, adverse effects during pregnancy, and even serious cardiovascular adverse events (see Chap. 4 in this book for more details), ignoring these effects as potential disadvantages seriously undermines the argument. Indeed, EDM's requirements of training and licensing for CE users actually protect the interests and rights of the disadvantaged, who are frequently targeted by promotional practices of the industry that promise social and other advantages by use of their product (consider yet again the example of *Rebel Without a Cause* and the chain-smoking hero it depicts). That is why I find the argument that EDM would be rejected based on the principle of fair opportunity unconvincing.

However, some might think that the principle of autonomy should take precedence over justice, and that my argument for different regulatory responses fails to respect autonomy sufficiently.

7.9 On the Respect for Autonomy and the Epistemic Position of Regulators

Flanigan (2013) argues that even amphetamine and instant-release forms of methylphenidate should be legally available because (1) prohibitions of recreational drugs are disrespectful to users, (2) even addicts are sufficiently autonomous with respect to their choice to use drugs, and (3) regulators are not in the best epistemic position to judge whether the risks of amphetamine and instant-release forms of methylphenidate warrant prohibition. She insists that even though some drugs reliably undermine users' life plans and autonomous capacities, policy makers should nevertheless permit drug use because the appropriate response to the value of autonomy is to respect individual choice, and not to promote an 'ideal' of autonomy.

Flanigan also claims that my argument rests on a dubious characterization of the psychology of addiction, as some empirical evidence suggests that addicts are autonomous when they choose to use drugs. Finally, she claims that it is illiberal to permit or prohibit drugs based on a judgment about whether the potential benefits to the user justify the risks users face. Her view is that consumers have the authority to decide for themselves whether it is worth it to risk their health for the sake of pharmacological benefits, either for enhancement or recreation.

Even though I agree with Flanigan that autonomy is a very important principle in liberal democracies, and that the state needs to respect it, I disagree with her view on what such respect entails. Consider the following analogy: Many individuals might find explosives helpful or fascinating for their rational life plans, and explosives could be used responsibly. However, the threat of irresponsible use of such objects is such a danger to others that the state is justified in restricting possession of explosives by members of the general public. As it is reasonable to endorse a system in which people are not allowed to walk around with explosives, it is perfectly rational and in accordance with autonomy to limit availability of amphetamines.

A substance for which there is overwhelming empirical proof that it can impair cognitive and volitional capacities (the very presuppositions of autonomy), and cause aggression and erratic and violent behavior (Miller 2002), is not (and should not be) subject only to the authority of the consumer. However, in both Chaps. 3 and 4, I do note the need to respect even some of the more self-destructive wishes of competent adults. Indeed, it will be remembered that I have concluded that when a person does voluntarily and autonomously choose to consume amphetamines with full knowledge of their addictive properties and harmful physiological and social consequences, the society would only be legitimate in punishing the producers and distributors of these drugs, but not the users.

7.10 On Some First Steps Toward Responsible Use of CE

Greely (2013) starts his objection on the positive side: by praising my scrutiny of the safety and efficacy of methylphenidate and amphetamine. He finds particularly important the analysis of the differences between the extended-release and the instant-release forms of medications. Greely notes the necessity to go beyond summarizing the existing research to pointing out what further research would be useful. He points out that we know very little about the effects of long-term use, either regular or sporadic, of these stimulant drugs on healthy adults. Greely notes that it is also necessary to have a discussion of mechanisms to assure that unbiased scientists would produce relevant research on various cognitive-enhancing drugs.

After pointing out strengths of my analysis and potential areas of improvement, Greely criticizes my dismissal of the proposal for a "Regulatory Authority for Cognitive Enhancement" (RACE). Even though Greely agrees with me that creating a new statutory regulatory body like RACE is difficult and expensive, and that it might not conform to the 1971 UN Convention on Psychotropic Substances, he thinks that those flaws might be overcome. His proposal is similar to mine—the use of an existing agency, similar to the Food and Drug Administration (FDA) in the United States. Greely contends that FDA might be better placed not only to make comparisons between safety aspects of different drugs, but also to forecast problems to which intermediate regulatory models need to respond, and examine, carefully, the match between those problems and available solutions. He points out that the focus of my analysis was on legitimate or appropriate policies, not about feasible politics.

Greely concludes that more needs to be said about foreseeable ways in which non-ideal conditions of politics may affect even ideal recommendations. He does note the constraints of space and considers my limited case analysis as a good start for further discussion by experts, governments, and the public.

Indeed, Greely is right that more needs to be said on a range of topics, from feasible politics via other possible models to discussing further avenues of unbiased research on CE drugs. However, I do feel the need to clarify one issue. Even though he believes that RACE could be amended to outmatch EDM as a regulatory solution, such an argument has not been provided yet. Furthermore, the precise reasons why EDM is preferable to RACE go beyond a limited case analysis of two stimulant drugs and delve deep into the values of a democratic society (see, e.g., Chaps. 1 and 2 of this book). For example, the questions of justice and accumulation of power could potentially cast serious doubts on RACE. The provisions of EDM, on the other hand, are precisely motivated by these "ideal" concerns of respect for individual decision making and benefiting the least advantaged, as opposed to merely efficiently regulating the market.

7.11 On the Need to Include All Stakeholder Perspectives

Forlini et al. (2013) agree with my case by case approach, as applied to Ritalin and Adderall, but they argue that public policies on CE should not only be based on an assessment of benefits and harms of the substances but also be informed by evidence on the perceptions and views of the groups that are affected by CE (i.e., stakeholders). They point out that assessment of evidence may be influenced by the experts' personal perceptions, experiences, and values and that the assessment of benefits and harms may well differ depending on whether a person is directly affected by a policy recommendation or not. Forlini and colleagues emphasize the fact that the involvement of stakeholders affected by policies has been incorporated as a requirement for the assessment of the quality of clinical guidelines. They add that empirical research indicates that stakeholder views on the appropriateness of a policy rest on more fundamental values that could be promoted or jeopardized by a liberal policy. Forlini and colleagues also point out that their research showed that effort put into an academic performance, which is linked to the authenticity of persons, is an important value to consider for stakeholders. The further point of criticism Forlini and colleagues raise is that my argument did not distinguish between requirements of moral acceptability (and here they include efficacy, safety, and respect for autonomy) and the requirements of moral praiseworthiness (among which they include self-realization, moral growth, and justice). They conclude that research on stakeholder perspectives shows that the criterion of moral praiseworthiness captures concerns about the impact that cognitive enhancement might have on values beyond safety and efficacy.

Forlini and colleagues rightly point out aspects that need to be included to increase democratic legitimacy of any proposal regarding regulation of CE. Indeed, the analysis of social harms of stimulants should include the points of view of general practitioners who prescribe stimulants, pharmacists who dispense them, and those who are using stimulants, whether to deal with a medical condition or as a study aid (see Dubljević 2018 for a longer argument). Not only would that provide valuable information, but it would also further promote values of the democratic society. As I noted earlier, I also agree with the need to go beyond safety, efficacy, and autonomy—the questions of justice and not jeopardizing self-realization (e.g., the right to an open future) are indeed important. However, I strongly disagree with their conclusion regarding authenticity and moral growth as criteria for public policy. As I argued elsewhere (Dubljević 2012a), these considerations could be included at the level of personal or even institutional (e.g., university) choice, but not at the level of society.

References

Beauchamp, T.L., and J.F. Childress. 2009. *Principles of biomedical ethics* (6th ed.). New York, NY: Oxford University Press.

Bigelow, B.C. 2006. *UXL encyclopedia of drugs and addictive substances*. Detroit: UXL/Thompson Gale.

Brewer, C.D., and H. De Grote. 2013. Regulating methylphenidate: Enhancing cognition and social inequality. *American Journal of Bioethics* 13 (7): 47–49.

Brunoni, A.R., et al. 2011. Clinical research with transcranial direct current stimulation (tDCS): Challenges and future directions. *Brain Stimulation*. https://doi.org/10.1016/j.brs.2011.03.002.

Dubljević, V. 2012a. Principles of justice as the basis for public policy on psycho-pharmacological cognitive enhancement. *Law, Innovation and Technology* 4 (1): 67–83.

Dubljević, V. 2012b. Toward a legitimate public policy on cognition-enhancement drugs. *American Journal of Bioethics – Neuroscience* 3 (3): 29–33.

Dubljević, V. 2014. Response to open peer commentaries on "prohibition or coffee-shops: Regulation of amphetamine and methylphenidate for enhancement use by healthy adults". *American Journal of Bioethics* 14 (1): W1–W8. https://doi.org/10.1080/15265161.2014.862417.

Dubljević, V. 2016a. Enhancement with modafinil: Benefiting or harming the society? In *Cognitive enhancement: Ethical and policy implications in international perspectives*, ed. F. Jotterand and V. Dubljević. New York: Oxford University Press.

Dubljević, V. 2016b. Autonomy is political, pragmatic and post-metaphysical: A reply to open peer commentaries on 'autonomy in neuroethics'. *AJOB – Neuroscience* 7 (4), W1–W3.

Dubljević, V. 2018. Toward an improved Multi-Criteria Drug Harm Assessment process and evidence-based drug policies. *Frontiers in Pharmacology*. https://doi.org/10.3389/fphar.2018.00898.

Faulmüller, N., et al. 2013. The indirect psychological costs of cognitive enhancement. *American Journal of Bioethics* 13 (7): 45–47.

Flanigan, J. 2013. An argument for permitting amphetamines and instant release methylphenidate. *American Journal of Bioethics* 13 (7): 49–51.

Forlini, C., et al. 2013. How research on stakeholder perspectives can inform policy on cognitive enhancement. *American Journal of Bioethics* 13 (7): 41–43.

Greely, H. 2013. Some first steps toward responsible use of cognitive-enhancing drugs by the healthy. *American Journal of Bioethics* 13 (7): 39–41.

Hall, W., et al. 2013. Constraints on regulatory options for putatively cognitive enhancing drugs. *American Journal of Bioethics* 13 (7): 35–37.

Iuculcano, T., and R. Cohen Kadosh. 2013. The mental cost of cognitive enhancement. *The Journal of Neuroscience* 33 (10): 4482–4486.

Iversen, L. 2008. *Speed, ecstasy, ritalin: The science of amphetamines*. Oxford: Oxford University Press.

LaBuzetta, J.N. 2013. Moving beyond methylphenidate and amphetamine: The ethics of a better 'smart drug'. *American Journal of Bioethics* 13 (7): 43–45.

Levy, N. 2013. There may be costs to failing to enhance as well as enhancing. *American Journal of Bioethics* 13 (7): 38–39.

Mill, J.S. 1859. *On liberty*. Available online at: http://www.bartleby.com/130/. Accessed 23 Apr 2011.

Miller, R.L. 2002. *Encyclopedia of addictive drugs*. London: Greenwood Press.

Nutt, D., L.A. King, W. Saulsbury, and C. Blakemore. 2007. Development of a rational scale to assess the harm of drugs of potential misuse. *The Lancet* 369 (9566): 1047–1053.

Oxford Dictionary. 2013. Conservative. http://oxforddictionaries.com/definition/english/conservative. Accessed 5 Oct 2013.

Sandberg, A. 2013. Enhancement policy and the value of information. *American Journal of Bioethics* 13 (7): 34–35.

Savulescu, J. 2013. A liberal consequentialist approach to regulation of cognitive enhancers. *American Journal of Bioethics* 13 (7): 53–55.

United Nations [UN]. 1971. *Convention on psychotropic substances*. www.unodc.org/pdf/convention_1971_en.pdf. Accessed 3 Apr 2013.

van der Eijk, Y. 2013. A blurry line between metaphysical free will and autonomy in addiction. *AJOB Neuroscience* 4 (4): 58–60.

Conclusion

Neuroscience offers a new and fresh perspective into ethics: both in terms of providing knowledge that qualifies ethical principles such as autonomy, and in providing technologies such as stimulant drugs and brain stimulation devices that demand ethical scrutiny. The use of these technologies can produce great injustice, or benefits to humanity, and I have argued that we should consider policies that would bring about the latter. The three parts of this book recommend a constructive engagement with neuroscience and technology, in conceptual, empirical and political perspectives that demand scrutiny.

However, as is the case with any conceptual analysis, a practical objection often arises. Namely, is the ethical and philosophical discussion of any pragmatic use? Indeed, in the context of regulation of cognitive enhancement drugs and devices, it could be the case that no society ever implements, or even considers implementing the Economic Disincentives Model, the Service Provider Model, or any other regulatory policy for cognitive enhancers discussed in this book. This global objection can also be reiterated in any particular context. For instance, would the German health system ever allow for a liberal approach to cognitive enhancement, or would the American political system with active lobbying from the pharma industry ever allow a model that employs heavy taxation to be enacted?[1]

It has to be said that no amount of philosophical replies can completely silence partial interests or bring conservative jurisdictions to consider less punitive approaches. But moral and political considerations can succeed in a more modest task. Namely, the current drug control regime is in a world-wide flux, and there is a new impetus to regulate the global market for neurostimulation devices. To the extent that some societies consider different alternatives to regulation, or merely that there are voices expressing concern about the inefficiency of current approa-

[1]I am grateful to Catrin Misselhorn (personal communication) for the first point, and to Greely (2013) for the second.

V. Dubljević, *Neuroethics, Justice and Autonomy: Public Reason in the Cognitive Enhancement Debate*, The International Library of Ethics, Law and Technology 19,
https://doi.org/10.1007/978-3-030-13643-7

ches, the arguments developed in this book will hopefully help in the public debates to come.

These arguments might be objected to in many ways. In part III, multiple critical comments both on the redefinition of the concept of autonomy and on its implementation in public policy have been reviewed. However, providing arguments and counter-arguments even in a public debate in which the citizens are engaged sometimes does not make any difference—politicians and institutions are set in their ways, or they just shrug off arguments, with a "so what?" attitude. Routine can certainly be much more powerful than public debate, and sometimes debates stay confined in the realm of the academic dispute, and have no resolution or immediate impact in the real world.[2] For instance, one of the normative conclusions of the argument from justice was that no public funds should go towards enhancement research, and yet the military (which is funded from tax payer's money) routinely funds enhancement research, and there is apparently no prospect of this changing. Similarly, the provisions of the Economic Disincentives Model, and the normative argument supporting it were that funds generated from taxation of enhancement should go to the least well off, and not, say, to military spending. This leads to a more general worry: the whole book appears to have a very narrow understanding of politics, presenting it as equivalent to policy and regulation.[3]

Again, these are all valid concerns. Even though this book tried to canvass the distinctly political approach in neuroethics (based on Rawlsian liberal political philosophy), the major focus was not on the political process, institutions and major actors that have a stake in the current social agenda, but on the practice of public reason, constrained by the forceless force of the better argument. Given time, a consensus (or a near-consensus) on an issue will lead politicians in liberal democratic societies to consider such an issue, include it in their agenda, and thus shape institutions. As mentioned above, drug policies are now in a state of flux. The inertia of the drug control system was tremendous—the routine still affects many people even in states in which, say, use of certain drugs such as cannabis is decriminalized. The Netherlands was very much a fore-runner in introducing the coffee-shop model, and for decades was the only society in which use of certain drugs was tolerated. The discussion on how it came to pass that similar models are now being carefully explored in certain jurisdictions in the United States and different countries around the world (e.g., Uruguay and Canada) is certainly of interest, but this is not something that this book deals with. Similarly, a discussion on the exact agents of change in the Netherlands, that have led to the coffee-shop policy being devised and implemented in the first place is certainly worthy and political science literature rightly addresses such issues. However, the institutions, agents and regulations in question start from democratic deliberation of citizens

[2]Indeed, the issues discussed in this book seem to defy the standard distinction between Ideal and Non-Ideal Theory in political philosophy, and signal the need to do more work in the uncharted territory between them. I am grateful to Tim Hinton for pointing this out.

[3]Special thanks to an anonymous reviewer of this book for pressing me on this point.

who have concerns about the implementation of fundamental democratic principles of justice and autonomy. Indeed, in the case of drug policy changes in certain states in the U.S., this was done by citizens concerned about their rights and autonomy via initiative and referenda activities (see McBride and Terry-McElrath 2016). This book cannot provide a full review of all matters political as they pertain to derivative uses of principle of autonomy (e.g., informed consent procedures) in specific settings or to the agenda setting in the matter of uses of stimulant drugs and stimulation devices.

What it can and does do is provide suggestions, in a preliminary way and in accordance with the Socratic spirit of philosophical midwifery, about how justice and autonomy might figure in the public reasoning of citizens of democratic societies. Great philosophers of past ages focused on problems that were troubling their societies based on their take of important ideas and lofty ideals that guide humanity. I hope that I have shown, on the examples of cognitive enhancement and neurobiology of addiction, that philosophy in the present day and age needs to address not only the deepest challenges of our social existence, but also some empirical details and political constraints. If I have been successful, at least neuroethics and neurophilosophy will no longer ignore the political aspect, even though it might not (or should not) be the only relevant perspective.

References

Greely, H. 2013. Some first steps toward responsible use of cognitive-enhancing drugs by the healthy. *American Journal of Bioethics* 13 (7): 39–41.

McBride, D.C., and Y. Terry-McElrath. 2016. Drug policy in the United States. In *The handbook of drugs and society*, ed. H.H. Brownstein, 574–593. New York, NY: Wiley.

Bibliography

Academy of Medical Sciences [AMA]. 2008. Brain science, addiction and drugs. An Academy of Medical Sciences working group report chaired by Professor Sir Gabriel Horn FRS FRCP. http://www.acmedsci.ac.uk/p99puid126.html. Accessed 5 Mar 2013.

Agar, N. 2005. *Liberal eugenics: In defence of human enhancement*. Malden, MA: Blackwell.

Albert, D.J. 1966. The effects of polarizing currents on the consolidation of learning. *Neuropsychologia* 4: 65–77.

Anderson, J. 2008. Neuro-prosthetics, the extended mind, and respect for persons with disability. In *The contingent nature of life. Bioethics and limits of human existence*, ed. M. Düwell, C. Rehmann-Sutter, and D. Mieth, 259–274. Dordrecht: Springer.

Aristotle. 1909. *Nichomachean ethics*, trans. L.H.G. Greenwood. Cambridge: Cambridge University Press.

Ashcroft, R.E. 2003. American biofutures: Ideology and utopia in the Fukuyama/Stock debate. *Journal of Medical Ethics* 29 (1): 59–62.

Audi, R. 2000. *Religious commitment and secular reason*. New York: Cambridge University Press.

Bainbridge, W., and M. Roco (eds.). 2006. *Managing nano-bio-info-cogno innovations: Converging Technologies in society*. Dordrecht: Springer.

Ballon, J.S., and D. Feifel. 2006. A systematic review of modafinil: Potential clinical uses and mechanisms of action. *Journal of Clinical Psychiatry* 67 (4): 554–566.

Barger, L.K., S.W.L. Lockley, S.M. Rajaratnam, and C.P. Landrigan. 2009. Neurobehavioral, health, and safety consequences associated with shift work in safety-sensitive professions. *Current Neurology and Neuroscience Reports* 9 (2): 155–164.

Bastuji, H., and M. Jouvet. 1988. Successful treatment of idiopathic hypersomnia and narcolepsy with modafinil. *Progress in Neuro-Psychopharmacology and Biological Psychiatry* 12: 695–700.

Bell, S.K., J.C. Lucke, and W.D. Hall. 2012. Lessons for enhancement from the history of cocaine and amphetamine use. *American Journal of Bioethics – Neuroscience* 3 (2): 24–29.

Bell, S., B. Partridge, J. Lucke, and W. Hall. 2013. Australian university students' attitudes towards the acceptability and regulation of pharmaceuticals to improve academic performance. *Neuroethics* 6 (1): 197–205.

Benhabib, S. 1986. The generalized and the concrete other: The Kohlberg-Gilligan controversy and feminist theory. *Praxis International* 5 (4): 402–424.

Birbaumer, N. 2005. Nur das Denken bleibt: Neuroethik des Eingeschlossen-Seins. In *Neurowissenschaften und Menschenbild*, ed. E.M. Engels and E. Hildt, 77–94. Paderborn: Mentis.

V. Dubljević, *Neuroethics, Justice and Autonomy: Public Reason in the Cognitive Enhancement Debate*, The International Library of Ethics, Law and Technology 19,
https://doi.org/10.1007/978-3-030-13643-7

Blank, R. 2014. Regulating cognitive enhancement technologies: Policy options and problems. In *Cognitive enhancement: Ethical and policy implications in international perspectives*, ed. F. Jotterand and V. Dubljević, under contract with Oxford University Press.

Bostrom, N. 2008a. Dignity and enhancement. In *Human dignity and bioethics: Essays commissioned by the President's Council on bioethics*, 173–207. Washington.

Bostrom, N. 2008b. Smart policy: Cognitive enhancement in the public interest. In *Reshaping the human condition. Exploring human enhancement*, 29–36. The Hague: Rathenau Institute.

Bostrom, N., and R. Roache. 2007. Ethical issues in human enhancement. http://www.nickbostrom.com/ethics/human-enhancement.pdf. Accessed 10 Nov 2012.

Button, M. 2005. Arendt, Rawls, and public reason. *Social Theory and Practice* 31 (2).

Cahill, M. 2005. The ethical consequences of modafinil use. *Penn Bioethics Journal* 1 (1): 1–3.

Caldwell, J.A., N.K. Smythe, J.L. Caldwell, et al. 1999. The effects of modafinil on aviator performance during 40 hours of continuous wakefulness: A UH-60 helicopter simulator study. *U.S. Army Aeromedical Research Laboratory*, Report No. 99-17.

Campa, R. 2008. Pure science and the posthuman future. *Journal of Evolution and Technology* 19 (1): 28–34.

Caulkins, J.P., P. Reuter, and C. Coulson. 2011. Basing drug scheduling decisions on scientific ranking of harmfulness: False promise from false premises. *Addiction* 106: 1886–1890.

Clark, C.S. 1995. Job stress. *CQ Researcher* 5: 681–704. http://library.cqpress.com/cqresearcher/document.php?id=cqresrre1995080400&type=hitlist&num=0. Accessed 11 Apr 2013.

Clark, A. 2003. *Natural-born cyborgs: Minds, technologies, and the future of human intelligence*. Oxford, New York: Oxford University Press.

Cohen, Joshua. 2003. For a democratic society. In *The Cambridge companion to Rawls*, ed. S. Freeman, 86–138. Cambridge: Cambridge University Press.

Cohen, Jon. 2013. Memory implants. *MIT Technology Review*. http://www.technologyreview.com/featuredstory/513681/memory-implants/#!. Accessed 23 Apr 2013.

Collins, D.J., and H.M. Lapsley. 1992. Drug abuse economics: Cost estimates and policy implications. *Drug and Alcohol Review* 11: 379–388.

CONTECS (ed.). 2008. *Converging technologies and their impact on the social sciences and humanities*. http://www.contecs.fraunhofer.de/images/files/contecs_report_complete.pdf. Accessed 11 Oct 2012.

Creutzfeldt, O.D., G.H. Fromm, and H. Kapp. 1962. Influence of transcortical d-c currents on cortical neuronal activity. *Experimental Neurology* 5: 436–452.

DAK. 2009. *Gesundheitsreport 2009. Analyse der Arbeitsunfähigkeitsdaten. Schwerpunktthema Doping am Arbeitsplatz*. Berlin: IGES.

Dancy, J. 2004. *Ethics without principles*. Oxford: Oxford University Press.

Daniels, N. 2011. Reflective equilibrium. In *The Stanford encyclopedia of philosophy*, ed. E.N. Zalta, Spring 2011 Edition. http://plato.stanford.edu/archives/spr2011/entries/reflective-equilibrium/ Accessed 11 Oct 2012.

Delgado, J. 1969. *Physical control of the mind: Toward a psycho-civilized society*. New York: Harper & Row.

Department of Justice. 2008. Biopharmaceutical company, Cephalon, to pay $425 million & enter plea to resolve allegations of off-label marketing. http://www.justice.gov/opa/pr/2008/September/08-civ-860.html. Accessed 30 Jan 2013.

Deroche-Gamonet, V., M. Darnaudéry, L. Bruins-Slot, F. Piat, M. Le Moal, and P.V. Piazza. 2002. Study of the addictive potential of modafinil in naïve and cocaine-experienced rats. *Psychopharmacology* 161: 387–395.

DeSantis, A.D., E.M. Webb, and S.M. Noar. 2009. Nonmedical ADHD stimulant use in fraternities. *Journal of Studies on Alcohol and Drugs* 70 (6): 952–954.

Dockery, C.A. 2013. The human experiment: How we won't win the rat race. What can we learn from brain stimulation in humans and rats about enhancing the functional neurobiology of higher cognitive functions. In *Cognitive enhancement: An interdisciplinary perspective*, ed. E. Hildt and A.G. Franke, 73–95. Dodrecht: Springer.

Dodge, T., K.J. Williams, M. Marzell, and R. Turrisi. 2012. Judging cheaters: Is substance misuse viewed similarly in the athletic and academic domains? *Psychology of Addictive Behaviors*, Advance online publication. https://doi.org/10.1037/a0027872.

Dreben, B. 2003. On Rawls and political liberalism. In *The Cambridge companion to Rawls*, ed. S. Freeman, 316–346. Cambridge: Cambridge University Press.

Dubljević, V. 2010. Rawls's formulation and reformulation of the Idea of Public Reason [Article in Serbo-Croatian]. *Pravni Život* 14: 737–758.

Dubljević, V. 2014. Cognitive enhancement: A glance at the future and ethical considerations. In *Cognitive enhancement*, ed. S. Knafo and C. Venero, under contract at Elsevier.

Dubljević, V., S. Sattler, and E. Racine. 2014b. Cognitive enhancement and academic misconduct: A study exploring their frequency and relationship. Ethics & Behavior 24(5): 408–420.

Dubljević, V., C. Venero, and S. Knafo. 2014b. What is cognitive enhancement? In *Cognitive enhancement*, ed. S. Knafo and C. Venero, under contract at Elsevier.

Düwell, M., D. Mieth, and C. Rehmann-Sutter. 2008. *The contingent nature of life: Bioethics and limits of human existence*. Dordrecht: Springer.

Estrada, A., A.M. Kelley, C.M. Webb, J.R. Athy, and J.S. Crowley. 2012. Modafinil as a replacement for dextroamphetamine for sustaining alertness in military helicopter pilots. *Aviation, Space, and Environmental Medicine* 83 (6): 556–564.

European Foundation for the Improvement of Living and Working Conditions [EFILWC]. 2009. *Second European quality of life survey*. Luxembourg: Office for Official Publications of the European Communities.

Evans, M. 2003. Public reason as liberal myth: Impartialist liberalism, judicial review and the cult of the constitution. *Journal of Transatlantic Studies* 1 (1): 8–25.

Farah, M.J. 2011. Neuroscience and neuroethics in the 21th century. In *Oxford handbook of neuroethics*, ed. J. Illes, 761–782. Oxford: Oxford University Press.

Farrely, C. 1999. Public reason, neutrality and civic virtues. *Ratio Juris* 12 (1): 11–25.

Fernandes De Oliveira, N. 2000. Critique of public reason revisited: Kant as arbiter between Rawls and Habermas. *Veritas* 45 (4): 583–606.

Finnis, J. 1980. *Natural law and natural rights*, 1st ed. *Clarendon law series*. Oxford: Clarendon Press.

Finnis, J. 1983. *Fundamentals of ethics*. Oxford: Oxford University Press.

Finnis, J. 1999. Natural law and the ethics of discourse. *Ratio Juris* 12 (4): 354–373.

Fischer, B., and P. Kendall. 2011. Nutt et. al'.s harm scales for drugs—Room for improvement but better policy based on science with limitations than no science at all. *Addiction* 106: 1891–1892.

Fischman, M.W., and R.C. Folting. 1991. Utility of subjective-effects measurements in assessing abuse liability of drugs in humans. *British Journal of Addiction* 86: 1563–1570.

Fitz, N.S., and P. Reiner. In Press. The perils of DIY cognitive enhancement: Taking minimally-invasive transcranial direct current stimulation seriously. In *The stimulated brain*, ed. R. Cohen-Kadosh et al. Oxford: Oxford University Press.

Forlini, C., and E. Racine. 2009. Autonomy and coercion in academic "cognitive enhancement" using methylphenidate: Perspectives of key stakeholders. *Neuroethics* 2 (3): 163–177.

Franke, A., and K. Lieb. 2010. Pharmakologisches Neuroenhancement und „Hirndoping": Chancen und Risiken. *Bundesgesundheitsblatt* 53: 853–860.

Franke, A.G., C. Bonertz, M. Christmann, M. Huss, A. Fellgiebel, E. Hildt, et al. 2010. Non-medical use of prescription stimulants and illicit use of stimulants for cognitive enhancement in pupils and students in Germany. *Pharmacopsychiatry* 44 (2): 60–66.

Freeman, S.R. 2003. *The Cambridge companion to Rawls*. Cambridge, U.K., New York: Cambridge University Press.

Fregni, F., et al. 2005. Transcranial direct current stimulation. *British Journal of Psychiatry* 186: 446–447.

Friedman, M. 2000. John Rawls and the political coercion of unreasonable people. In *The idea of political liberalism*, ed. V. Davion and C. Wolf, 16–33. Lanham: Rowman & Littlefield.

Fukuyama, F. 2003. *Our posthuman future: Consequences of the biotechnology revolution*, 1st ed. New York: Picador.

Gaus, G.F. 1997. Reason, justification and consensus—Why democracy can't have it all. In *Deliberative democracy: Essays on reason and politics*, 205–242. London: MIT Press.

George, R.P. 2006. Public morality, public reason. *First Things*, www.firstthings.com/article.php3?id_article=5344. Accessed 28 Sept 2011.

Gerrard, P., and R. Malcolm. 2007. Mechanisms of modafinil: A review of current research. *Neuropsychiatric Disease and Treatment* 3 (3): 349–364.

Gibson, J.J. 1986. *The ecological approach to visual perception*. Hillsdale, NJ: Lawrence Erlbaum Associates Inc.

Gilbert, F. 2013. Nano-bionic devices for the purpose of cognitive enhancement: Toward a preliminary ethical framework. In *Cognitive enhancement: An interdisciplinary perspective*, ed. E. Hildt and A.G. Franke, 125–138. Dodrecht: Springer.

Giordano, J.J., and B. Gordijn. 2010. *Scientific and philosophical perspectives in neuroethics*. Cambridge, UK, New York: Cambridge University Press.

Guleyupoglu, B., P. Schestatsky, D. Edwards, F. Fregni, and M. Bikson. 2013. Classification of methods in transcranial electrical stimulation (tES) and evolving strategy from historical approaches to contemporary innovations. *Journal of Neuroscience Methods* 219: 297–311.

Habermas, J. 1976. *Zur Rekonstruktion des Historischen Materialismus*. Frankfurt am Main: Suhrkamp.

Habermas, J. 1995. Reconciliation through the public use of reason: Remarks on John Rawls's political liberalism. *The Journal of Philosophy* 92 (3): 109–131.

Habermas, J. 2001. *Die Zukunft der menschlichen Natur: Auf dem Weg zu einer liberalen Eugenik?*. Frankfurt am Main: Suhrkamp.

Habermas, J. 2007a. *Moral consciousness and communicative action*. Cambridge: Polity Press.

Habermas, J. 2007b. The language game of responsible agency and the problem of free will: How can epistemic dualism be reconciled with ontological monism? *Philosophical Explorations* 10 (1): 13–50.

Hamani, C., M.P. McAndrew, M. Cohn, M. Oh, D. Zumsteg, C.M. Shapiro, R.A. Wennberg, and A. Lozano. 2008. Memory enhancement induced by hypothalamic/fornix deep brain stimulation. *Annals of Neurology* 63 (1): 119–123.

Haraway, D. 1991. *Simians, cyborgs and women: The reinvention of nature*. New York: Routledge.

Harsanyi, J.C. 1975. Review: Can the maximin principle serve as a basis for morality? A critique of John Rawls's theory. *The American Political Science Review* 69 (2): 594–606.

Health and Safety Executive [HSE]. 2006. *Managing shift work*. London: HSE Books.

Heersmink, R. 2013. Embodied tools, cognitive tools and brain-computer interfaces. *Neuroethics* 6: 207–219.

Heymann, J., A. Earle, and J. Hayes. 2007. *The work, family, and equity index: How does the United States measure up?* Montreal, QC and Boston, MA: Institute for Health and Social Policy—Project on Global Working Families.

Hughes, J. 2006. Human enhancement and the emergent technopolitics of the 21st century. In *Managing nano-bio-info-cogno innovations: Converging technologies in society*, ed. W. Bainbridge and M. Roco, 285–307. Dordrecht: Springer.

Ilinskiĭ, O.B., V.P. Lebedev, A.B. Savchenko, A.I. Solovyeva, and S.E. Spevak. 1987. Effect of transcranial non-invasive stimulation of the antinociceptive structures of the brain on processes of repair (Article in Russian). *Fiziologicheskiĭ zhurnal SSSR imeni I. M. Sechenova* 73 (2): 223–229.

JASON (Program for Office of Defense Research and Engineering). 2008. *Human performance*. JASON Advisory group for Dep. of Defence, Pentagon, JSR-07-625. www.fas.org/irp/agency/dod/jason/human.pdf. Accessed 11 Apr 2011.

Jex, S.M. 2002. *Organizational psychology: A scientist-practitioner approach*. New York: Wiley.

Kalant, H. 2010. Drug classification: Science, politics, both or neither? *Addiction* 105: 1146–1149.

Kanal, M.G., C. Ozkan, O. Doganavsargil, and M. Eryilmaz. 2012. Late onset mania possibly related to modafinil use: A case report. *Bulletin of Clinical Psychopharmacology* 22 (1): 71–74.

Kass, L. 2002. *Life, liberty, and the defense of dignity*. San Francisco: Encounter Books.

Kasselheim, A.S., J.A. Myers, D.H. Solomon, W.C. Winkelmayer, R. Levin, and J. Avorn. 2012. The prevalence and cost of unapproved uses of top-selling orphan drugs. *PLoS ONE* 7 (2): e31894. https://doi.org/10.1371/journal.pone.0031894.

Kim, D. 2012. Practical use and risk of modafinil, a novel waking drug. *Environmental Health and Toxicology*. http://dx.doi.org/10.5620/eht.2012.27.e2012007. Accessed 23 Mar 2012.

Klerman, G. 1972. Psychotropic hedonism vs. pharmacological calvinism. *The Hastings Center Report* 2 (4): 1–3.

Knutsson, A., N. Hammar, and B. Karlsson. 2004. Shift workers' mortality scrutinized. *Chronobiology International* 21 (6): 1049–1053.

Kolker, R. 2013. The real limitless drug isn't just for lifehackers anymore. *New York Magazine*, 31 March 2013. http://nymag.com/news/intelligencer/modafinil-2013-4/. Accessed 2 Mar 2013.

Lagarde, D., D. Batejet, P. Van Beers, D. Sarafian, and S. Pradella. 1995. Interest of modafinil, a new psychostimulant, during a sixty-hour sleep deprivation experiment. *Fundamental Clinical Pharmacology* 9 (3): 271–279.

Larmore, C. 2003. Public reason. In *The Cambridge companion to Rawls*, ed. S. Freeman, 368–393. Cambridge: Cambridge University Press.

Lee, S., D. McCann, and J.C. Messenger. 2007. *Working time around the world: Trends in working hours, laws and policies in a global comparative perspective*. NewYork: Routledge.

Lennard, N. 2009. One in ten takes drugs to study. *Varsity* 693: 1.

Lynch, Z. 2006. Neuropolicy (2005–2035): Converging technologies enables neurotechnology creating new ethical dilemmas. In *Managing nano-bio-info-cogno innovations: Converging technologies in society*, ed. W. Bainbridge and M. Roco, 173–192. Dordrecht: Springer.

Lynch, Z., and B. Laursen. 2009. *The neuro revolution: How the brain science is changing our world*. New York: St. Martin's Press.

Mariani, J.D., and C.L. Hart. 2005. Psychosis associated with modafinil and shift work. *American Journal of Psychiatry* 162 (10): 18.

Maritain, J. 1951. *Man and the state*. Chicago: University of Chicago Press.

Marx, K. 1875. Critique of the Gotha program. http://www.marxists.org/archive/marx/works/1875/gotha/ch01.htm. Accessed 3 June 2013.

McCabe, D.L., and L.K. Trevino. 1993. Academic dishonesty: Honor codes and other contextual influences. *The Journal of Higher Education* 64 (5): 522–538.

McCabe, D.L., L.K. Trevino, and K.D. Butterfield. 2001. Cheating in academic institutions: A decade of research. *Ethics & Behavior* 11 (3): 219–232.

McCabe, D.L., L.K. Trevino, and K.D. Butterfield. 2002. Honor codes and other contextual influences on academic integrity: A replication and extension to modified honor code settings. *Research in Higher Education* 43 (3): 357–378.

McGee, E.M. 2010. Toward regulating human enhancement technologies. *American Journal of Bioethics - Neuroscience* 1 (2): 49–50.

Medical Bureau of Roadside Safety [MBRS]. 2012. *Report on roadside drug testing and equipment and related matters*. Dublin: Medical Bureau of Roadside Safety.

Metzinger, T., and E. Hildt. 2011. Cognitive enhancement. In *Oxford handbook of neuroethics*, ed. J. Illes and B. Sahakian, 245–264. Oxford: Oxford University Press.

Mikhaĭlova, N.G., and K.I. Sarkisova. 1977. Effect of excitation summation during interaction of self-stimulation zones (Article in Russian). *Zhurnal vysshei nervnoi deiatelnosti imeni I P Pavlova* 27 (5): 1020–1029.

Minzenberg, M.J., and C.S. Carter. 2008. Modafinil: A review of neurochemical actions and effects on cognition. *Neuropsychopharmacology* 33: 1477–1502.

Mohamed, A.D. 2012. Modafinil has the potential for addiction. *American Journal of Bioethics - Neuroscience* 3 (2): 36–38.

Mohamed, A.D., and B.J. Sahakian. 2012. The ethics of elective psychopharmacology. *International Journal of Neuropsychopharmacology* 15: 559–571.

Moon, D.J. 2003. Rawls and Habermas on public reason: Human rights and global justice. *Annual Review of Political Science* 6: 257–274.

Morse, S.J. 2012. Legal regulation of addictive substances and addiction. In *Addiction neuroethics: The ethics of addiction neuroscience research and treatment*, ed. A. Carter, W. Hall, and J. Illes, 261–276. London: Academic Press.

Müller, S., and H. Walter. 2010. Reviewing autonomy: Implications of the neurosciences and the free will debate for the principle of respect for the patient's autonomy. *Cambridge Quarterly of Healthcare Ethics* 19: 205–217.

Nnodim, P. 2004. Public reason as a form of normative and political justification: A study on Rawls's idea of public reason and Kant's notion of the use of public reason in *What is enlightenment?. South African Journal of Philosophy* 23 (2).

Nordmann, A. 2007. If and then: A critique of speculative nanoethics. *Nanoethics* 1: 31–46.

Norman, C., and M. Berger. 2008. Neuroenhancement: Status quo and perspectives. *European Archives of Psychiatry and Clinical Neuroscience* 258: 110–114.

Nussbaum, M.C. 2003. Rawls and feminism. In *The Cambridge companion to Rawls*, ed. S. Freeman, 488–520. Cambridge: Cambridge University Press.

Nutt, D. 2011. Let not the best be the enemy of the good. *Addiction* 106: 1892–1893.

Nutt, D.J., L.A. King, and L.D. Phillips. 2010. Drug harms in the UK: A multicriteria decision analysis. *Lancet* 376: 1558–1565.

Okin, S.M. 1994. Political liberalism, justice, and gender. *Ethics* 105: 23–43.

Parens, E. 2005. Authenticity and ambivalence: Toward understanding the enhancement debate. *Hastings Center Report* 35 (3): 34–41.

Parrott, A.C. 2007. Drug-related harm: A complex and difficult concept to scale. *Human Psychopharmacology Clinical and Experimental* 22: 423–425.

Partridge, B., J. Lucke, and W. Hall. 2012. A comparison of attitudes toward cognitive enhancement and legalized doping in sport in a community sample of Australian adults. *American Journal of Bioethics - Primary Research* 3 (4): 81–86.

Patterson, R. 2004. Reviewing public reason: A critique of Rawls's political liberalism and the idea of public reason. *Deakin Law Review* 9 (2).

Pepperell, R. 2003. *The posthuman condition: Consciousness beyond the brain*. Bristol: Intellect.

Quong, J. 2011. *Liberalism without perfection*. Oxford: Oxford University Press.

Racine, E., and V. Dubljević. In Press. Neuroethics: Neuroscience and society. *Oxford Handbooks Online – Philosophy.*

Racine, E., and J. Illes. 2008. Neuroethics. In *Cambridge textbook of bioethics*, ed. P. Singer and A. Viens, 495–503. Cambridge: Cambridge University Press.

Rawls, J. 1997. The idea of public reason revisited. *University of Chicago Law Review* 64 (3): 765–807.

Rawls, J. 2002. *The law of peoples with the idea of public reason revisited*. Cambridge: Harvard University Press.

Rawls, J. 2003. *Lectures on the history of moral philosophy*. London: Harvard University Press.

Rawls, J. 2007. *Lectures on the history of political philosophy*. Cambridge, MA: Belknap Press of Harvard Univ. Press.

Redfearn, J.W.T., O.C.J. Lippold, and R. Costain. 1964. Preliminary account of the clinical effects of polarizing the brain in certain psychiatric disorders. *The British Journal of Psychiatry* 110: 773–785.

Repantis, D., P. Schlattmann, O. Laisney, and I. Heuser. 2010. Modafinil and methylphenidate for neuroenhancement in healthy individuals: A systematic review. *Pharmacological Research*. https://doi.org/10.1016/j.phrs.2010.04.002.

Riedel, W. 2008. Psychopharmaceutical cognition enhancement. In *Reshaping the human condition: Exploring human enhancement*, ed. L. Zonneveld, H. Dijstelbloem, and D. Ringoir, 115–122. The Hague: Rathenau Institute.

Roads and Traffic Authority [RTA]. 2008. *Roadside drug testing*. Sydney: New South Wales Centre for Road Safety. http://www.rta.nsw.gov.au/heavyvehicles/downloads/hv_drug_testing_dl1.html. Accessed 02 May 2013.

Robert Koch Institute [RKI]. 2012. Pharmacological neuroenhancement. *GBE KOmpakt* 3 (3). http://edoc.rki.de/series/gbe-kompakt/2010-1/PDF/1_eng.pdf. Accessed 2 May 2013.

Roberts Skerrett, K. 2005. Political liberalism and the idea of public reason: A response to Jeffrey Stout's *Democracy and Tradition*. *Social Theory and Practice* 31 (2).

Rusinov, V.S. 1965. Polarization of the motor region of the cerebral cortex and the motor dominant by direct current anode (Article in Russian). *Zhurnal vysshei nervnoi deiatelnosti imeni I P Pavlova* 15 (2): 217–228.

Sahakian, B., and S. Morein-Zamir. 2007. Professor's little helper. *Nature* 450 (7173): 1157–1159.

Sandberg, A., and J. Savulescu. 2008. Neuroenhancement of love and marriage: The chemicals between us. *Neuroethics* 1 (1): 31–44.

Sattler, S., and C. Wiegel. 2013. Test anxiety and cognitive enhancement: The influence of students' worries on their use of performance-enhancing drugs. *Substance Use & Misuse* 48 (3): 220–232.

Scanlon, T.M. 2003. Rawls on justification. In *The Cambridge companion to Rawls*, ed. S. Freeman, 139–167. Cambridge: Cambridge University Press.

Scheffler, S. 2003. Rawls and utilitarianism. In *The Cambridge companion to Rawls*, ed. S. Freeman, 426–459. Cambridge University Press.

Schermer, M. 2008. On the argument that enhancement is "cheating". *Journal of Medical Ethics* 34 (2): 85–88.

Schermer, M. 2013. Reducing, restoring or enhancing autonomy with neuromodulation techniques. In *Free will and the brain: Neuroscientific, philosophical, and legal perspectives*, ed. W. Glannon (In press). Cambridge: Cambridge University Press.

Sen, A. 2009. *The idea of justice*. Cambridge, MA: Belknap Press of Harvard University Press.

Sententia, W. 2006. Cognitive enhancement and the neuroethics of memory drugs. In *Managing nano-bio-info-cogno innovations: Converging technologies in society*, ed. W. Bainbridge and M. Roco, 153–172. Dordrecht: Springer.

Sharwood, L.N., J. Elkington, L. Meuleners, R. Ivers, S. Boufous, and M. Stevenson. 2013. Use of caffeinated substances and risk of crashes in long distance drivers of commercial vehicles: case-control study. *British Medical Journal* 346, E-pub ahead of print 20 Mar 2013. http://dx.doi.org/10.1136/bmj.f1140.

Shortall, M. 2009. *Human Rights and Moral Reasoning—A comparative investigation by way of three theorists and their respective traditions of inquiry: John Finnis, Ronald Dworkin and Jürgen Habermas*. Rome: Gregorian & Biblical Press.

Smith, J.L. 2004. The 44-hour day. *The Telegraph* (6 Jan 2004) http://www.telegraph.co.uk/health/3304214/The-44-hour-day.html. Accessed 11 Apr 2013.

Sterba, J.P. 1999. Reconciling public reason and religious values. *Social Theory and Practice* 25 (1).

Tannenbaum, J. 2012. The promise and peril of the pharmacological enhancer modafinil. *Bioethics*. https://doi.org/10.1111/bioe.12008.

Trade Commodities. 2011. The most common commodities to trade. http://www.tradecommodities.co.uk/commodities/. Accessed 3 Apr 2013.

Utz, K.S., et al. 2010. Electrified minds: Transcranial direct current stimulation (tDCS) and galvanic vestibular stimulation (GVS) as methods of non-invasive brain stimulation in neuropsychology—A review of current data and future implications. *Neuropsychologia* 48: 2789–2810.

Volkow, N.D., J.S. Fowler, J. Logan, D. Alexoff, W. Zhu, et al. 2009. Effects of modafinil on dopamine and dopamine transporters in the male human brain. *The Journal of the American Medical Association* 301: 1148–1154.

Vrecko, S. 2013. Just how cognitive is 'cognitive enhancement?' *American Journal of Bioethics – Neuroscience* 4 (1): 4–12.

Walzer, M. 1983. *Spheres of justice*. New York, NY: Basic Books.
Williams, E., et al. *Human performance*. JASON (Program for Office of Defense Research and Engineering) (No. JSR-07-625). JASON Advisory group for Dep. of Defence, Pentagon. www.fas.org/irp/agency/dod/jason/human.pdf. Accessed 3 Apr 2013.
Wolpe, P.R. 2004. Neuroethics. In *The encyclopedia of bioethics*, ed. S.G. Post, 1894–1898. New York: MacMillan.
Yudko, E., H.V. Hall, and S.B. McPherson. 2003. *Methamphetamine abuse: Clinical and forensic aspects*. New York: CRC Press.
Zonneveld, L., H. Dijstelbloem, and D. Ringoir (eds.). 2008. *Reshaping the human condition: Exploring human enhancement*. The Hague: Rathenau Institute.

Index

V. Dubljević, *Neuroethics, Justice and Autonomy: Public Reason in the Cognitive Enhancement Debate*, The International Library of Ethics, Law and Technology 19,
https://doi.org/10.1007/978-3-030-13643-7

The manufacturer's authorised representative in the EU is Springer Nature Customer Service Centre GmbH, Europaplatz 3, 69115 Heidelberg, Germany. If you have any concerns regarding our products, please contact ProductSafety@springernature.com

Printed and bound by CPI Group (UK) Ltd, Croydon, CR0 4YY
05/10/2026
02228154-0005